T0100387

THE ILLUSIONIST BRAIN

Joan Brossa, *Alfa*, 1986

If he were not a writer, in moments of exhilaration he would be a guerrilla,
and in moments of peacefulness, a magician. Being a poet involves both.
—Joan Brossa, *Vivarium* (1972)

THE ILLUSIONIST BRAIN

The Neuroscience of Magic

Jordi Camí and Luis M. Martínez

Translated by Eduardo Aparicio

PRINCETON UNIVERSITY PRESS
PRINCETON AND OXFORD

Original Spanish edition, *El cerebro ilusionista: La neurociencia detrás de la magia*, 2020, RBA Books

Published by Princeton University Press
41 William Street, Princeton, New Jersey 08540
99 Banbury Road, Oxford OX2 6JX

press.princeton.edu

Library of Congress Cataloging-in-Publication Data

Names: Camí, Jordi, 1952– author. | Martínez, Luis M., 1969– author. | Aparicio, Eduardo, translator.
Title: The illusionist brain : the neuroscience of magic / Jordi Camí and Luis M. Martínez ; translated by Eduardo Aparicio.
Other titles: Cerebro ilusionista. English
Description: Princeton : Princeton University Press, [2022] | Includes bibliographical references and index.
Identifiers: LCCN 2021051899 (print) | LCCN 2021051900 (ebook) | ISBN 9780691208442 (hardback) | ISBN 9780691239156 (ebook)
Subjects: LCSH: Optical illusions. | Magic tricks. | Neurosciences. | BISAC: SCIENCE / Life Sciences / Neuroscience | PSYCHOLOGY / Cognitive Neuroscience & Cognitive Neuropsychology
Classification: LCC QP356 .C3613 2022 (print) | LCC QP356 (ebook) | DDC 612.8—dc23/eng/20211221
LC record available at https://lccn.loc.gov/2021051899
LC ebook record available at https://lccn.loc.gov/2021051900

British Library Cataloging-in-Publication Data is available

Editorial: Hallie Stebbins and Kristen Hop
Production Editorial: Mark Bellis
Text Design: Karl Spurzem
Jacket Design: Chris Ferrante
Production: Erin Suydam
Publicity: Sara Henning-Stout and Kate Farquhar-Thomson
Copyeditor: Cynthia Buck

Jacket illustration by Natalya Balnova, Marlena Agency

Frontispiece image used with permission from Fundació Joan Brosssa

This book has been composed in Arno

Printed on acid-free paper. ∞

Printed in the United States of America

10 9 8 7 6 5 4 3 2 1

Contents

THE ILLUSIONIST BRAIN

CHAPTER 1

The Art and Science of the Impossible

For many of us, magic is a portal to the days of childhood. Our earliest experiences with the art of the impossible are often experiences of much fanfare, of circuses with a magnificent and varied cast of artists, or costumed magicians making cards appear from bare hands. Even if you have seen only the occasional magic show, no doubt the following scenario will feel somewhat familiar to you, as the experience of magic in many ways unites its spectators, even as magic itself stands apart from all other art forms.

A magician onstage shows her audience several seemingly ordinary pieces of wood and, in an instant, assembles them into a cube-shaped box. She places the box on a table, nonchalantly opens it, and from it removes a beautiful bouquet of purple and yellow flowers. Her audience applauds this unexpected appearance. The magician then invites an audience member onstage, asking them to thoroughly examine the inside and outside of the box, to verify that there was nothing special about it. The magician again closes the box and immediately opens it again to remove countless colored handkerchiefs. The audience erupts into a loud and synchronized round of applause. Finally, the magician invites the volunteer to examine the box once more, to ensure that it is indeed empty, and this time asks the volunteer to close it after she is done. The magician opens the box and a pair of beautiful white doves emerge and briefly take flight before returning to her hands. The audience starts again to applaud, many looking at each other in disbelief, unable to believe what they have just witnessed.

How did the magician accomplish this feat—the art of the impossible? Magic is an active dialogue between the magician and her audience, but the language is not one that we understand. In the following pages, we will decipher a little of that language, shedding light on how magic works—and how it manages to fool our brains.

The Art of the Impossible

Juan Tamariz, an influential Spanish magician who enjoys enormous international prestige, has said that "the art of magic must have the purpose of raising the quota of happiness in the world, in others and in ourselves."[1] Magic is indeed a unique art, capable of making spectators feel that something has happened that is impossible, that defies the laws of nature. Tamariz believes that the outcome of a magic trick must be unexpected, seemingly impossible, and fascinating.[2] It is *unexpected* in that expectations are shattered, especially in those magic tricks in which no element of the exposition anticipates what will happen at the end. The outcome is *impossible* in that magic outcomes contradict logic or the laws of nature, and it is *fascinating* in that the final effect of good magic is dazzling and extraordinary. For Tamariz, only magic combines impossibility with fascination, while other activities, such as certain acrobatics, may be fascinating but are not perceived as impossible. The essential and specific quality of a magic trick's climax is the "mystery" of the impossible, as it combines a confrontation with the unknowable with the "mental shock" of seeming impossibility.[3]

How does magic—illusionism, the art of creating impossible effects that violate our expectations, tricks that conclude with the apparent transgression of natural laws—work?

In this book, we explore the cognitive processes behind the art of magic, an ancient artistic activity that, after centuries of trial and error, has accumulated an important repository of wisdom regarding its techniques.

A magic trick always begins with a demonstration, a story, or a proposition that concludes with a seemingly impossible, fascinating, and unexpected result. As spectators, we are captivated by the disparity between what we assume will happen at the end of the trick and what we finally observe happening. These outcomes are tremendously provocative. They contradict our hypotheses and make us doubt everything we have learned. They are a cognitive dissonance in and of themselves.

Why are our brains taken by surprise? Why does magic collide with all our mental schemes and often break them? As we will see in the following pages, magic works because it takes advantage of the limitations of the brain's normal processing. Magicians understand how our brains process information. They know where the brain's weaknesses are, and they know how to design tricks that capitalize on those weaknesses by manipulating our attention and perception so that we overlook important details, or making us see or hear things that are not there.

The human brain is a very advanced organ. Its capabilities are highly perfected and adapted to our environment and lifestyle, so much so that we usually are not aware of its limitations, both physical and metabolic. These limitations, however, are very real: every moment our brains receive an enormous amount of information through our senses, far more information than we are aware of. Yet another limitation is speed: the transmission of information between neurons is relatively slow and must overcome several bottlenecks throughout the brain—points at which one part of the brain circuit holds back the potential transmission and processing capacity.

Through evolution, the brain has overcome these limitations by developing extraordinarily effective strategies to process everyday sensory information. For instance, the brain may build an illusion of a continuous experience where there isn't one, or it may make inferences based on limited data, as when we recognize a person from afar simply by the way they walk.

In computer terms, we could say that magic, surprisingly, has learned to hack into some of these strategies using the brain's "back doors." Magic reveals the tricks and automations that characterize the functioning of our brains, interfering and, above all, playing with our unconscious processing.

Magic most often enters the brain via the sense of sight, because a human being is an extremely visual animal, so much so that more than one-third of the cerebral cortex is dedicated to processing the information captured by the retinas. Through the process of paying attention and using our short-term memories, the brain filters and selects only that information it considers useful at any given moment. This information is incomplete, captured in a fragmented way in both space and time, yet the brain must use its complex resources to create the illusion of continuous reality. Thus, when we perceive a scene in front of us, our brains do not see an accurate reflection of reality but rather infer it. To do this, our brains rely heavily on the prior knowledge we have accumulated in our long-term memories.

For example, when we observe a bird flying between the trees of a forest, appearing and disappearing as it flies past each tree, our brains do not think that different birds appear in sequence from behind each tree. We know perfectly well that it is always the same bird. It seems obvious that we would perceive the situation correctly, but it actually is not. We have no physical proof that it is the same bird; we could, in fact, imagine the opposite situation: different birds emerging from behind each tree in perfect synchrony. Our previous experience as individuals and as a species, however, accumulated through evolution, allows us to come up with the most plausible explanation.

When magicians trick us, they are interfering with all of the brain's strategies for inferring reality. They take advantage of the fact that sometimes we are blind to certain changes in our environment, and they manipulate our perception and memory in such a way that, even though we look, we do not see what is happening.

In the presentation of a magic trick, a distinction is made between the "external life" of the effect, which is what the audience sees and enjoys, and the hidden "internal life." Happening secretly, the "internal life" is what makes the whole magic experience possible. It is a parallel reality, completely separate from the guiding thread that the magician spins for spectators to follow.

Contrast can be a central mechanism in the external life of a magic trick. Usually, the human brain more easily processes information with some contrast—that is, with changes, differences. For example, there is a very high contrast between this white book page and the black letters on it. If we reduced the black letters to a very light gray, it would be harder to perceive them, as there would be less contrast against the white page. Sometimes magicians want to create contrast by capturing an audience's attention—such as when they pull a rabbit out of a hat—so that the audience "sees" something. At other times, however, when magicians are performing maneuvers that must go completely unnoticed by the audience, they avoid at all costs the provocation of contrast.

As we will see in this book, magic can also manipulate our memories, condition us, and influence our intuitive decisions, all without our realizing it. Magic tricks deceive us because they are presented with a logic and a naturalness that hardly seem suspicious. Everything is predictable until the surprising outcome that shatters our expectations. This surprising outcome, the climax of the magic act, is the crucial point of the trick. It is very difficult to master, as it requires that the magician challenge our capacity to infer and anticipate, processes that are not under our conscious control. This is why we say that magic speaks to, challenges, and deceives our unconscious brain.

Though this deception can be extraordinarily difficult to achieve, the results are worth the effort. The outcome of a magic trick triggers multiple emotions and intellectual reactions and is an experience unlike any other. In this way, magic is also fundamentally an art, a performance, and as such, it is presented in playful contexts. Perhaps its inherent artistry explains why, surprisingly, magic has been so little studied by science over the centuries.

In the following pages, we show that magic and science have much to offer each other. This book is fundamentally an exploration of neuroscience, through the lens of magic. We explain how magic works through a discussion of the brain's normal, day-to-day processes—which themselves may sometimes seem like acts of magic. Through our discussion, it will become clear that there is much about the intersection of magic and neuroscience that remains unexplored territory. We hope to show the great potential not only for neuroscience to shed light on how magic works, but for the world of magic to open new and unexpected doors of knowledge for neuroscience.

Where We Will Go in This Book

Although magic uses many techniques and devices drawn from scientific disciplines as varied as mathematics, physics (including optics), mechanics, electronics, chemistry, and new materials, in this book we will focus on cognition.

When we refer to "cognitive processes," we mean those tasks or operations that the brain executes continuously to process the information we receive from the environment: attention, perception, memories, emotions, decision-making, reasoning, planning, problem-solving, and learning (focusing here on those processes that the magician usually controls or manipulates). It is through our cognitive processes that we create, analyze, and interact with reality, all the while relying on our prior experience and knowledge. Cognitive processes thus allow us to be flexible and to adapt our behavior almost immediately based on the changes and demands imposed by the different situations of everyday life.

When we discuss "magic tricks," we are focusing exclusively on the mechanisms of magic tricks that provoke the "illusion of impossibility"—the ones that audiences consider impossible because what happens at the end goes against the laws of nature. We do not cover the techniques used by "psychics" or any other practitioner of a method of divination; their universe of knowledge is different from—but not alien to—the procedures, resources, and methods used by magicians.*

Magic has its own schools, experts, and centuries of accumulated experience. Beyond its deceits, magic is a scenic art that combines resources from theater and other sources to achieve successful effects, always at the service of a surprising outcome. After centuries of tests and empirical trials, today's magic is the result of a wisdom accumulated over time, based on experience and the perfecting of an immense catalog of materials and methods that magicians have created and baptized with their own names or unique characteristics.

*We use the word "magician" and not "illusionist" to refer to the person who performs magic tricks, although the two terms are synonymous. Some theorists of magic propose a distinction between the two, but we have ignored it here for practical reasons. Similarly, we use the word "magician" to refer to the person who performs magic, regardless of the genre.

In the past, the world of magic was responsible for discovering and validating these techniques. Today neuroscience wants to learn from this wisdom. The American magician Persi Diaconis, a scientist and professor of statistics, has verified that original contributions from magic have helped open new pathways of knowledge in the mathematical fields of cryptography and the analysis of DNA sequences.[4] Our aim is to follow the lead of mathematics and facilitate an equally fruitful dialogue between magic and neuroscience.

The Grammar of Magic

To perform good magic, magicians rely on solid principles based on experience, most of which respond to cognitive processes. During the second half of the twentieth century, some theorists of magic, like the Spanish magicians Arturo de Ascanio and Juan Tamariz or the American Darwin Ortiz, developed authentic "grammars" of their language. In this book, we often refer to concepts coined by Arturo de Ascanio. Ascanio was born in 1929, and in the 1950s, after meeting the great Dutch magician Fred Kaps, he created a vast work on magic that he continued to build and elaborate on until his death in 1997. One of our goals has been to interpret and "translate" this language coined by magicians into concepts that cognitive neuroscientists can use to explain how the brain works. As the following pages demonstrate, we are convinced that exploring how magic works can bring new perspectives to neuroscience.

Your Journey with Us

In part I, we lay the foundation for understanding the neuroscience of magic. Chapters 2 and 3 present a simplified model of the structure and function of the brain, with special emphasis on the visual pathway, because magic enters through the sense of sight.

In part II, we examine the different cognitive processes involved in magic tricks. In chapter 4, we describe how the brain creates an illusion of continuity to compensate for the fact that we capture external information in a fractured way in both space and time. We'll see that magic takes advantage of this phenomenon in multiple ways. In chapter 5, we describe the key concept of contrast: magicians can either avoid or provoke contrast as a tool for attention control.

In chapter 6, we turn to attention, one of the brain processes that magic has learned to control with great precision. Through our attention, we continuously filter and select from the enormous amount of information we receive. Chapter 7 explores the creative world of perception, arguing that perceiving is literally a process of interpretation.

The neuroscience behind magic does not stop with information processing. As we will see in chapter 8, magicians are actually able to manipulate our memories during the few minutes that a magic trick lasts.

Chapter 9 looks at how magicians can condition us and take advantage of the multiple mechanisms of the unconscious brain. Moreover, as chapter 10 shows, magicians, unbeknownst to us, also know how to induce certain responses and decisions.

Part III of the book opens with chapter 11, a reflection on the magic experience and different audience reactions to it. We should note that all discussions of the magic experience in this book refer mainly to the Western culture with which we are familiar. In other cultures, the illusionist or magic experience may be interpreted differently: some may conclude that the magician possesses divine powers, while others may react to inexplicable effects with fear and aggressiveness. In this book, however, we do not delve into these other perspectives.*

To close the book, chapter 12 recognizes the pioneering research efforts on magic that were made at the end of the nineteenth century and details how much more is still to be done.

* An eloquent example of the importance of cultural context in magic was provided by the magician Jean Eugène Robert-Houdin in 1856. Having retired, he was required by the French colonial government of Napoleon III to travel to Algeria and demonstrate "his magic powers" before the Marabouts, the spiritual and religious leaders of the Arab tribes. The local inhabitants' belief that the Marabouts were endowed with magic powers posed a challenge to the authority of the colonists, and Napoleon wanted Robert-Houdin to help neutralize the influence of the Arab leaders by showing that French magic was stronger. Robert-Houdin toured and performed several shows in Algeria, and among his tricks (or "powers") was the lifting of a box secretly attached to the ground at will by an electromagnetic charge; see Prevos, *Perspectives on Magic*.

We hope we have conveyed in this first chapter that magic is able to seduce us with its effects because of the way our brains understand the world around us. Knowing how the brain works and understanding the cognitive processes involved in magic effects can help us develop a full appreciation of magic—and help theoreticians of magic create the best possible magic.

At the same time, the empirical knowledge that magic has accumulated over time is a valuable source of knowledge for neuroscience. Though some techniques used in magic tricks correspond to well-known mechanisms in the field of cognitive neuroscience, scientists do not yet understand the processes that underlie other magic techniques. These techniques therefore offer very attractive research opportunities. Many neuroscientists believe that artists in general—and magic is no exception—have intuitively discovered, after years of trial and error, how the brain works and how it interprets the world. Artists use this knowledge to enhance the impact of their work.

Jorge Wagensberg expressed that idea when he said: "The least banal relationship between science and art occurs when an artist offers scientific intuition to a scientist or when a scientist offers artistic insight to an artist."[5]

Finally, this book is a recognition of the scientific foundations of magic and of those who practice it honestly and with good intent, as opposed to those who use its methods for illegitimate purposes or to make the public believe that they are endowed with supernatural powers, as some psychics do. Partway into the twenty-first century, we believe that there is no artistic endeavor important enough to justify deceiving spectators. We do well enough constantly deceiving ourselves.

PART I

The Basics

CHAPTER 2

Living in Illusion

The Human Brain and the Visual Pathway

> What is life? An illusion, a shadow, a fiction. And the greatest good is trifling, for all life is but a dream, and dreams, mere dreams.
>
> —Pedro Calderón de la Barca, *Life Is a Dream* (1630)

We Live in Illusion

Human beings literally live in illusion. We interpret or "construct" our own reality from the information we receive from the universe around us. This is an automatic, unconscious process: we are not aware of our brains doing this, or of the strategies the brain uses to offer us such a comfortably predictable reality. It is difficult to accept that we only imagine everything we see and that we use limited information to systematically anticipate what will happen. If this depiction of our reality seems like science fiction, it is because our inferences, based on previous experience, are usually extraordinarily accurate and reliable, and it is these inferences that ultimately create our reality.

Our sense organs are mere instruments specialized in receiving external stimuli and transmitting them to the brain, the only organ that truly sees, touches, tastes, and hears. Indeed, though we detect external stimuli through the sense organs, light and colors are not in our

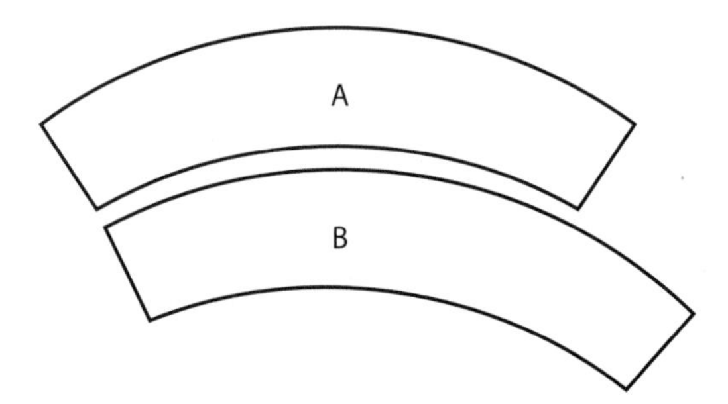

FIGURE 2.1. "Jastrow Illusion": This geometric illusion was attributed to Joseph Jastrow in 1891, although it was first described by the German psychologist Franz Müller-Lyer in 1889. Jastrow, born in Warsaw, was the first doctor of psychology in the United States, and he founded the Department of Psychology at the University of Wisconsin. Figures A and B are identical. This can be confirmed by copying them onto a piece of paper and superimposing the two figures. But it's inevitable that we perceive them as different. The bottom figure seems longer, probably because its longer edge appears next to the shorter edge of the top figure. When assessing their areas, we cannot help but consider the lengths of the lines that define them, and those differences influence our perception. In Jastrow's own words, "We judge relatively, even when we most desire to judge absolutely." From Jastrow, "Studies from the Laboratory of Experimental Psychology of the University of Wisconsin," 397.

eyes; touch is not in our hands; taste is not in our mouth; the noises of the city are not in our ears. Our brains are in charge of building reality. With incomplete data from the world around us, and using context and a catalog of its own, the brain fills in and completes images, builds emotions, reconstructs memories, shapes decisions, and categorizes people.

Thus, by adding details and filling in gaps, the brain arrives at guesses that are close and convincing. But these guesses are sometimes flawed, resulting in illusions (optical, visual, and cognitive), false memories, biases, and cognitive prejudices. It is precisely these illusions or biases that constitute proof that the brain works through highly elaborate and refined strategies based on very little information (see figure 2.1). The brain must use these strategies—these shortcuts—to function effectively. Taking shortcuts is easier and more efficient than reconstructing reality with total precision.[1]

The Brain, Its Cells, and Its Structure

As we mentioned in chapter 1, magic impacts the brain primarily through our visual system. A magician's sleight of hand enters through our retina and is transmitted to the visual system, where the brain makes sense of it, missing in the process a vital piece of information and eventually jumping to the wrong conclusion. How does this happen—how does the magician trick our visual system? To understand this, we'll need to understand a little of the anatomy of the brain.

The brain is a mass of nerve tissues composed of:

- The cerebral cortex: a gelatinous external layer about 3 to 5 millimeters thick
- The subcortical structures: the thalamus, hypothalamus, amygdala, and the basal ganglia, among other structures
- The cerebellum and the brain stem

The cerebral cortex is proportionally one of the densest parts of the brain, with about 16 billion neurons. The cortex has two symmetrical hemispheres, one left and one right, with a surface folded in on itself. Its ridges and grooves increase the cortical surface area to almost one square meter, without significantly altering its volume. Each hemisphere consists of four lobes: frontal, parietal, temporal, and occipital (see figure 2.2).

The frontal lobe houses in its most anterior part the prefrontal cortex, our large "executive center." In charge of supervising thoughts and decision-making, this is the lobe that enables us to plan. The parietal lobe is the great integrating center of motor and sensory stimuli. The temporal lobe houses and encloses structures such as the hippocampus and amygdala and is responsible for memory formation, spatial navigation, and management of emotions, among other functions. Finally, the

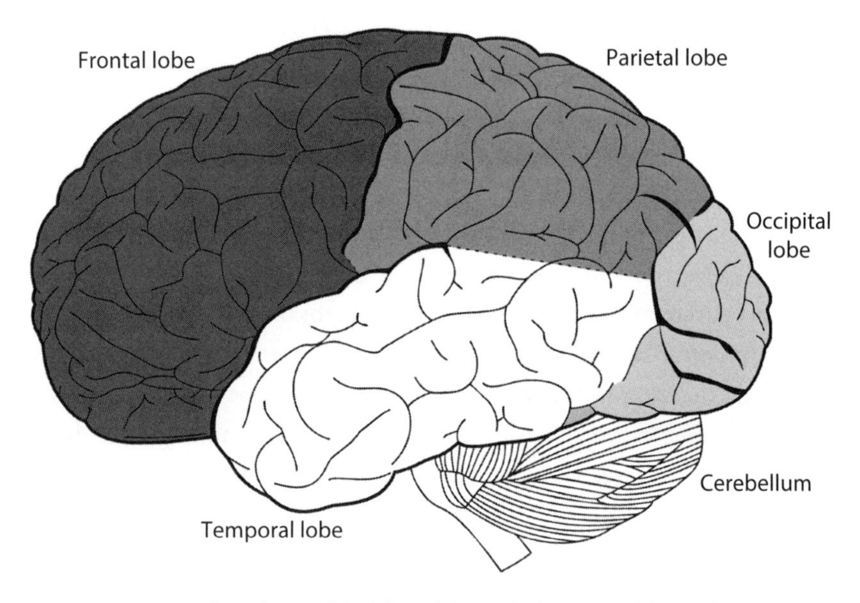

FIGURE 2.2. A lateral view of the lobes of the cerebral cortex and the cerebellum.

occipital lobe is the structure that integrates all the information we receive through the visual pathway.

The cerebral cortex envelops and relates to a multitude of subcortical structures, a group of diverse deep nuclei within the brain that includes, among others, the thalamus and hypothalamus of the diencephalon, the amygdala and associated limbic structures, and the basal ganglia. They play a pivotal role in cognitive, affective, and social functions in humans, participating in complex actions such as memory, emotion, bonding, and the sense of pleasure.

The cerebellum is a structure very similar to a miniature brain. It contains 70 percent of all the neurons in the brain and it participates in the control of movement, posture, and walking. The cerebellum is also a key organ for implicit learning and memory; some cognitive processes and social behaviors, both learned and automatic, are managed there. On our journey through the brain, we will explore all of these areas, but most of our focus will be on the cortex, as this area plays a leading role in human cognition.

Neurons

At the cellular level, there are three main types of cells in the brain: neurons, glial cells, and endothelial cells. Glial cells, among other functions, modulate interneural connections. Endothelial cells ensure brain vascularization, through which oxygen and nutrients reach the brain and waste is removed. Traditionally, neurons are considered the cells responsible for transmitting and managing both information that reaches the brain and information that is produced internally.

Neurons, of which there are an estimated 85 billion, have an overall structure that includes a cell body, an axon, and dendrites (see figure 2.3). Information is transmitted through an electrical impulse that travels along the axon of the neuron, which is usually lined with a myelin sheath that electrically isolates the axon membrane to more effectively transmit the impulse to the dendrites of another neuron. Dendrites, like the branches of a tree, are numerous in order to maximize the integration of information from different sources.

The electrical signal spreads across the surface of the neuronal membranes through complex ionic exchanges between the inside and outside of the membranes. Conduction speed can reach more than 500 kilometers per hour, and hundreds of electrical impulses can be produced per second.

While the transmission of information along a neuron is electrical in nature, between neurons it is generally of a chemical nature. Neurons communicate with each other through so-called synaptic connections at the gaps between neurons. Here the electrical signal transmitted along the membrane is converted into chemical information and results in the release of substances known as neurotransmitters. These include, for example, glutamate, noradrenaline, dopamine, serotonin, and gamma-aminobutyric acid. Neurotransmitters play such an essential role in neuron communication, and hence in brain function, that many psychoactive drugs act by simply modulating the action of these neurotransmitters. Most synapses are chemical, although there are also purely electrical synapses, whose transmission speed is much faster.

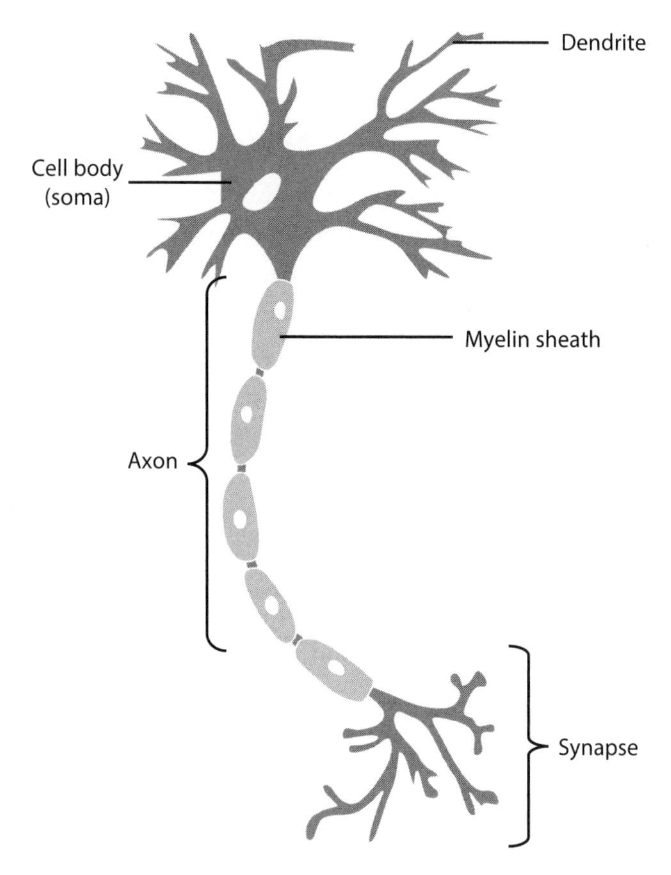

FIGURE 2.3. A stereotypical image of a neuron showing its different functional components.

Neural Networks

> Neurons that fire together wire together.
>
> —Donald Hebb, *The Organization of Behavior* (1949)

As early as 1949, Donald Hebb observed that when neurons are simultaneously activated, their connection is strengthened, indicating that they are encoding similar information. This model of joint organization is the basis for the concept of biological neural networks, functional connections between groups of neurons that work together and are dynamic, changing over time.

Thus, after learning or experiencing something, neurons create, modify, or undo their synaptic connections. The changes in these connections ultimately underpin the functional plasticity of the brain. While there are synaptic connections that are labile and ephemeral, others are stable and last over time, such as those that store learning or memories. Long-term memories—as we will see in chapter 8—require a degree of stability to ensure that some memories are never lost. Although it may seem contradictory, plasticity is also required to enable us to forget and to generate new memories.

In other words, strong, stable connections ensure the durability of memories; disconnections, on the other hand, promote forgetfulness and leave room for encoding new memories through new connections. In many cases, evoking memories leads to a reconstruction of the facts; we now believe that this reconstruction is expressed in a new synaptic configuration.

As expected, changes in the connections, disconnections, and adaptations of the synaptic connectivity are continuous: we wake up every day with a brain that is, anatomically or structurally speaking, different from that of the previous day. It is precisely while we sleep that we consolidate memories and, in turn, prune out a lot of information accumulated throughout the day.

This is why each person's brain is considered unique. The brain expresses the personal way in which we interact with the world, combining genetic dispositions with conformational variants derived from our personal experiences. Such is the particularity of each person's neural connections that some scientists even think that true identity lies not so much in our genes as in the connections of our neurons—that is, in our particular "wiring."

Whereas the cells of the skin's epidermis continually divide to ensure their replacement in the event of injury, neurons are rarely renewed: they are born and die with us. This permanence helps to preserve information—the experiences that end up shaping our identity—throughout our lives. Otherwise, with each division of neurons, we would start from scratch.

Beginning with the neuroscientist and Nobel Prize winner Santiago Ramón y Cajal, the human brain has been characterized as a huge interconnected network. Its connections and circuits are based on the relationships that neurons establish with each other by means of billions of synaptic connections, regardless of whether these connections are working or not.

As we have already seen, networks and circuits are dynamic entities. Their functional structures are formed by groups of neurons that work in a coordinated manner. They are modular in nature, because each network is dedicated to analyzing a specific type of information, and they are generally organized hierarchically, because that analysis is becoming more and more complex. There are, moreover, some very large networks, such as the visual, somatomotor, auditory, limbic, and frontoparietal networks, among others.

Some of these networks are more stable and dominant than others. One example is the "default mode network," also known as the "resting state." This is a network associated with resting states and higher cognitive functions, such as thoughts and internal mental states; it remains active even when no external stimulus or internal mental intent is being processed. This network tells us with certainty that the brain never rests.

The Visual Pathway

Approximately one-third of the brain is involved in processing vision, that is, interpreting color, detecting contours and movements, establishing depth and distance, determining the identity of objects, interpreting faces, and so on. Our eyes capture information from the outside world in the form of a shower of photons. In the retina, the photons are transformed into electrical signals that are transferred to the brain via the optic nerve. All of this electrical information is processed, for the most part, in the occipital lobe; more than a third of the cerebral cortex is involved in this process.

We live in a visually rich world, and processing it is the first challenge that our brains must face. We have more than 100 million photoreceptors in each of our retinas that receive a total of approximately 70 gigabytes of information per second—equivalent to watching seventy movies per second with images, dialogue, and soundtrack included! But clearly, it is not possible to manage so much information. That's why the brain has learned to select and filter what we see.

Ultimately, we actually collect and use very little information in the world. But even with this limited information, and thanks to a highly refined process of inference and decoding, we manage to create the illusion of reality to which we have been referring since the beginning of this book. Let us now look at the processes responsible for making this possible.

The retina covers the inner surface of the eye with millions of photoreceptors, a special type of neuronal receptor that specializes in capturing photons and transmitting that information through a chain of neurons in the eye that culminate in ganglion cells.

The axons of the ganglion neurons are the ones that form the optic nerve and are the only means of communication between the retina and the brain. The transmission of information at this point is the first major bottleneck that will condition how we later interpret any image. Of the 70 gigabytes that the eye receives per second, it is estimated that only one megabyte per second of information (1 MB/s, or 0.0014 percent of the total) is transmitted to the brain. Thus, the optic nerve has a transmission rate similar to that of an ethernet connection.

The optic nerve leaves the eye at a very specific point in the retina that lacks receptors, the so-called blind spot. This is a point where we should not see anything, but we do not even realize that the blind spot exists because the brain fills that gap by calculating an average of what we are seeing in the immediately neighboring regions. This phenomenon is known as "filling in" (see figure 2.4a).

In addition, there is "filling out," in which what we are seeing in a very specific and special area of the retina located near the blind spot called the fovea is extrapolated to peripheral regions of the visual field (see figure 2.4b). The fovea specializes in providing us with visual acuity.

Filling out occurs when the visual image of the fovea and the periphery are very similar. This illusion of uniformity—this phenomenon of "filling out" the peripheral stimuli with central information—is a process that has also been demonstrated in a wide range of visual characteristics, including shape, orientation, movement, luminance

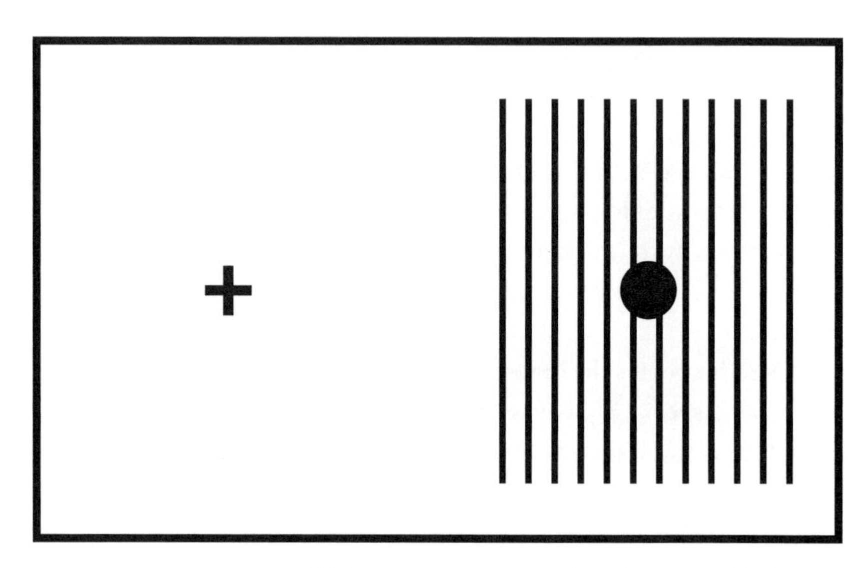

FIGURE 2.4a. Demonstration of "filling in": We see a cross and a circle. If we hold the book a couple of hand spans from our eyes, then close our left eye and, with the right, look directly at the cross, slowly moving the book back and forth, there comes a moment when the circle disappears and we perceive only the continuity of the striped pattern. When this happens, the image of the circle coincides exactly with the retina's blind spot.

(luminous flux that strikes, crosses, or emerges from a surface), pattern, and identity.[2]

The Photoreceptors: Cones and Rods

There are two types of photoreceptors in the human retina: cones and rods. Of the two, the most numerous (about 100 million) are the rods, which are distributed throughout the periphery. Cones are less numerous (about 6 million); they are concentrated in the fovea, which represents no more than 2 percent of the total surface of the retina and is the key to visual acuity.

Rods and cones have very different sensitivity to light and, in a way, are complementary. Rods are capable of responding to the arrival of a single photon on their membrane but adapt immediately when the intensity of light rises a little: they stop responding.

For this reason, rods enable us to see reasonably well at twilight, in gray scale, and when there is no color, but they are useless in daylight vision.

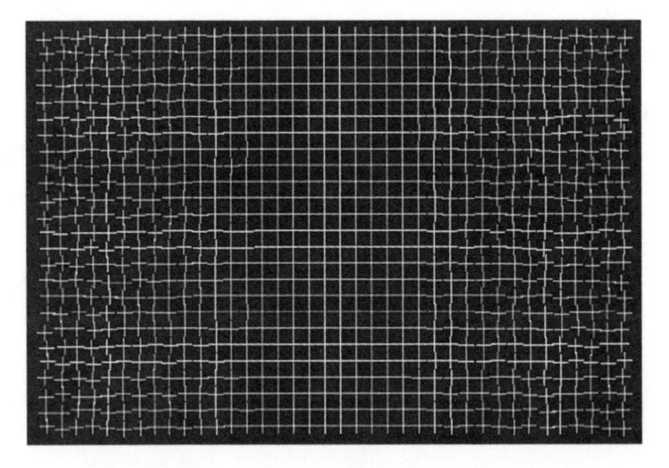

FIGURE 2.4b. Demonstration of "filling out": If we fix our eyes on the center of the image, after a few seconds the whole grid becomes a homogeneous image, just like the grid in the center. An unconscious process fills in the poor peripheral image with the detailed structure of the central image. Original image is from Ryota Kanai and is used with permission.

Cones, on the other hand, require a much higher luminous flux. They are activated only when thousands of photons arrive simultaneously.

There are three types of cones, and each has a preference for different wavelengths that correspond, approximately, to our experience of the colors red, green, and blue. The wide range of colors that we are able to perceive is built up by combining in different ratios the signals coming from these three types of cones.

Cones allow us to see clearly during the day; at night or in low-light conditions, they are inactive. This is why it is so dangerous to drive or walk on a poorly lit road at dawn or dusk. In such conditions, rods do not respond because they have adapted to the lighting of the moment, and the cones have not yet been activated because there is not enough light for them, resulting in dangerous areas of functional blindness.

This feature of rods and cones is an important reason why stage magicians sometimes present their effects in low-light conditions. Although inconspicuous, low lighting favors concealment and distorts the clarity with which we are accustomed to seeing things.

What the Brain Sees

As we have discussed, we select and attend to only a small part of the information we receive. Moreover, what the retina transmits to the brain is not a reliable reproduction. It is rather a type of information transformed to highlight edges, differences, and contrasts. Different parts of the brain will then use these elements to sequentially construct the image. Therefore, it is really the brain that sees, not the eyes.

In the remainder of this chapter, we will move from our overview of the nervous and visual systems to a discussion of how artists have helped us understand the basic principles of visual processing. We may think that neuroscience and art have little in common, but seeing the world through the eyes of an artist can enhance our understanding of the brain. After all, the work of a painter is in some ways not that different from the work of a neuroscientist (or a magician), and in many ways there is more that unites us than separates us.

For thousands of years, painters have tried to generate on a fixed, two-dimensional support—such as a stone wall or a canvas—images that resemble their perceptive experience of the world in which they live. To do so, they develop a personal language, their own grammar, based on a combination of patterns and forms, colors, and luminance. Neuroscientists, for their part, take the opposite route: they attempt to define the rules of the brain—that is, the internal grammar that allows the brain to reconstruct "a subjective reality" from the surrounding visual world. To do this, the brain, like the painter, relies solely on a succession of two-dimensional images continuously projected onto the retinas, as if they were a kind of canvas. Thus, painters and neuroscientists, both exploring clues related to perspective, color, form, movement, and contrast, seem to be looking at themselves in an imaginary mirror to better understand how we see the world.

One of the first things that artists in their work and scientists in the laboratory have discovered relates to the perception of form, though it has taken scientists forty thousand years longer to make this discovery than it took painters.

The Beginning of Art

The first artists, those who painted hunting scenes and animal bodies on the walls of caves, already realized that it was enough to draw the "edges" of an object to generate a very vivid perception of it. This is possible because the structure of the retina is such that its cells can detect the basic zones of an image in which the amount of light reflected or emitted suddenly produces a local change. We generically call these sudden changes "contrast," and they occur primarily along the edges of objects. As we explain later in the book, the concept of contrast will play a central role in the presentation of magic effects.

How does contrast appear in the brain? To start, we can say that the retina transforms every image into a drawing of simple lines. Over the centuries, artists have discovered that these types of drawings or basic sketches can be, and often are, more perceptively powerful than a faithful reproduction of the original image.

But why? Why can an image that provides less visual information be more suggestive than one richer in detail? The scientists Patrick Cavanagh and Vilayanur Ramachandran suggest that the simpler image is more suggestive because brain resources are limited and we cannot pay attention to all the visual details available in a complex image.[3] In fact, looking at a drawing like "Contour Simulation" (figure 2.5), we realize that all our attention is directed to the most relevant parts without being dispersed to less informative areas. This happens because, by always starting from filtered information, we actively "fill in" the rest of the image based on our previous experience and knowledge of the world. To do this, we generate associations between what we actually see, which is very limited, and our conceptualization of the object as represented in drawings that we have previously stored in our memory. Thus, by leaving so much room for the spectator's contribution, line drawings are usually even more suggestive than the real models that inspire them.

As we will see in chapter 4, this "filling in" process is a strategy that we constantly use to minimize various limitations of our brains.

FIGURE 2.5. "Contour Simulation": A simulation of how the receptor fields of the retina transform every image, like the one on the left, into a contour drawing, like the one on the right.

Color and Luminance

The peculiarities of our perception of a scene's color and light have also been widely exploited in the art world. Margaret Livingstone is a scientist who has dedicated her studies to explaining in detail how painters, especially since the Renaissance, have developed techniques that allow them to use our brains' preference for contrasting differences to generate the sensation of three dimensions in their paintings.[4]

For example, Leonardo da Vinci realized that placing colors reflecting a large amount of light, such as yellow, next to others reflecting less light, such as blue, generates high-contrast areas. The resulting illusion makes us believe that the latter are farther away in the visual scene than the former.

In her 2002 book *Vision and Art: The Biology of Seeing*, Dr. Livingstone explains artists' discovery that they can treat color and luminance independently on their canvases. It is possible to depict a scene in which there is contrast between the different colors but no difference in luminance. In fact, that is exactly what Claude Monet did in his painting *Impression, Sunrise*, which would give its name to the Impressionist movement (figure 2.6). As you can see, in this painting, the sun is very bright and seems to sparkle in a peculiar way in the morning sky. In the real world, the sun emits much more light than is reflected by the surrounding sky. In Monet's work, the sun, despite being a different color from the sky, reflects the same amount of light. It is therefore equally luminous and would not be perceived as brighter if the painting were transformed into a gray scale. It is precisely this lack of contrast in luminance that gives the painting its full appeal. As Picasso said, "Colors are only symbols, reality is to be found in lightness alone."[5]

FIGURE 2.6. Claude Monet, *Impression, Sunrise*: The bottom image is the gray-scale version of the painting. Reproduced with permission from Margaret Livingstone.

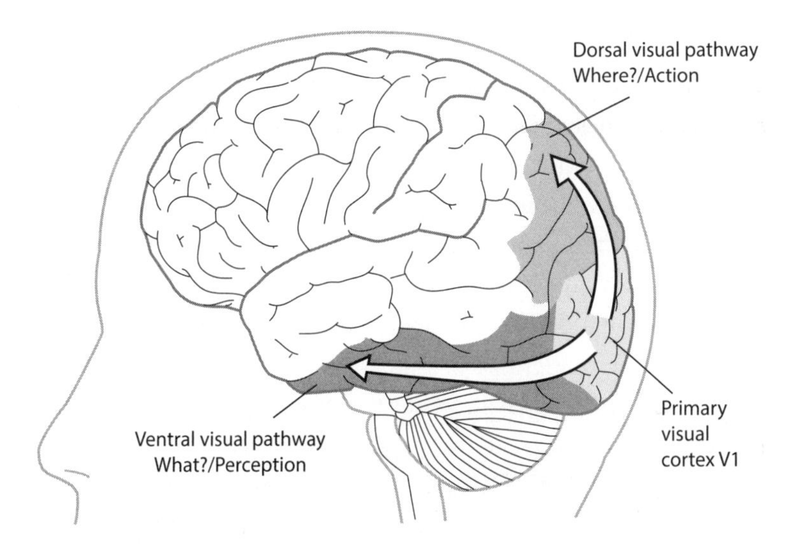

FIGURE 2.7. Functional organization of the visual system, indicating the ventral and dorsal pathways.

Color, or hue, and luminance can be artificially separated on a canvas because they are processed separately in the brain. Indeed, the visual system can be divided into two main pathways that differ not only in their location but also in their function (figure 2.7).

The "What" and "Where" Pathways

The most modern part of the visual pathway that, in evolutionary terms, we share only with other primates has its origin in the ganglion cells of the retina located mainly in the fovea (parvocellular), and it runs along the ventral area of the brain, in the occipital and temporal lobes. It has been called the "what" pathway because activity in this area correlates with our conscious recognition of objects that appear in a visual scene.

In other words, the "what" pathway is responsible for building the color, shape, and texture of objects, as well as the faces we meet in the street. Lesions in the temporal lobe can cause serious difficulties in the recognition of objects (agnosia) or faces (prosopagnosia). In addition, lesions in another area of this pathway called V4 can produce selective loss of color (achromatopsia), resulting in being able to see only black and white.

The other main pathway of the visual system originates in the more peripheral retinal ganglion cells; it continues dorsally into the cerebral cortex through the occipital and parietal lobes. A scene's depth, its three dimensions, the global and relative movement in it, and its organization are attributes analyzed in this pathway, called by many the "where" pathway. Injuries in this part of the visual system cause deficits in motor coordination that can lead to akinetopsia, also known as motion blindness.

The "where" pathway is the oldest component of our visual system in evolutionary terms. We share it with all mammals, and it is only sensitive to changes in luminance; its cellular components are "blind" to color. The "where," or dorsal, pathway is also faster but has a poorer spatial resolution; for this reason, the fine detail of an image is mainly analyzed in the ventral or "what" pathway.

With the understanding that visual processing on the two pathways is very different, many artists, especially from the nineteenth century onward, have been able to exploit this difference amazingly well in their work. The Impressionists, for example, realized that it does not matter what color is used to convey a difference in luminance, which, as we have shown, conveys the basic information of a scene. Therefore, they used completely unrealistic colors and luminance contrasts in their paintings to generate illusory sensations of brightness, depth, and movement.

Returning to Monet's painting, the sun in *Impression, Sunrise* (figure 2.6) is just as luminous as the background; it is only "seen" by the ventral pathway—the color-sensitive "what" pathway. The dorsal pathway—the "where" pathway responsible for perceiving the spatial

FIGURE 2.8. Renoir, *La Grenouillère* (top), Monet, *La Grenouillère* (bottom). Reproduced with permission from Margaret Livingstone.

FIGURE 2.9. Akiyoshi Kitaoka, *Rotating Snakes*. Reproduced with permission.

location of objects—does not see the sun (bottom of figure 2.7). Because our visual system is unable to accurately establish its position, the sun appears to be shining in the sky.[6]

The paintings of *La Grenouillère* (figure 2.8), painted simultaneously by Monet and Pierre Renoir, capture the atmosphere they experienced in this rest area on the banks of the Seine. However, Monet did a much better job of capturing the essence of the river water, its texture and movement. To do so, he used a characteristic color sequence: black, yellow, white, and blue. Black and white have higher luminance contrast than yellow and blue; an illusion of movement is created because the brain processes these differences in contrast at different speeds and therefore at different times.

The neuroscientist Akiyoshi Kitaoka independently discovered the same phenomenon (150 years later!) and was able to generate static images that produced his well-known visual illusions of movement by combining those same four colors (figure 2.9).[7]

The intention of the Impressionists was to capture the very essence of the image. They discovered that they could achieve this by blurring the edges and shapes of objects, using thick brushstrokes to leave only lower-spatial-frequency information, a lower resolution. Had they not done so, they would not have achieved such a captivating effect, because the high-frequency information provided by defined edges predominantly activates the ventral ("what") pathway and would have dominated the elusive perception generated in the dorsal ("where") pathway.

The Expression of Emotions and the Act of Seeing

Portraits by Impressionists can be tremendously emotional. It is not necessary to receive very detailed information in these portraits about an individual's facial expression to recognize or *feel* the person's mood and determine whether they are sad or happy, surprised or angry.

In real life, we recognize the mood of a fellow human being when the visual system sends a copy of the visual information down the dorsal pathway to a deep nucleus of the brain called the amygdala. As we will see in chapter 3, the amygdala is involved in processing emotions and receives mainly low-resolution visual information (such as the thick strokes of the Impressionists). The amygdala processes this information so quickly that we unconsciously feel the mood of our interlocutor long before we even recognize the person, and therefore long before we have analyzed all the visual information provided by their face. Interestingly, these clues used to appreciate a picture are the same clues used to perceive, quickly and effectively, the real world.[8]

However, this internal logic of the brain, its "alternative physics," in the words of Patrick Cavanagh, does not have to be realistic, so its rules are unpredictable. For instance, a shadow in a painting does not have to be black or dark gray, as in real life; it can be a very stunning color, such as orange or purple, as long as it has less luminance than the object that supposedly casts it. This alternative brain physics also explains our perception of other attributes, such as lighting, shading, perspective, and the apparent evenness of luminance and color.

The act of seeing, understood as our conscious vision of a scene, culminates when the information that has been filtered by our visual system reaches the association zones of the temporal cortex and generates intimate and precise links with our memories and predictions. Because our visual understanding of the world depends on our ability to generate associations, seeing becomes a highly creative process, and one with which we can interact very effectively. This is precisely what both artists and magicians do. In the following pages, we describe the how and why of these interactions: how visual information is attended to in the brain as it interacts with our memories and experience, and why magic tricks aimed at hacking into those interactions work.

CHAPTER 3

The Conception of Reality

We Are Our Memories

Perception of the Outside World

As we have discussed, the perception of reality that is built up in our brains is an illusory inner image that does not directly correspond to the world around us. Outside, colors do not exist; the brain constructs them through the action of electromagnetic waves that stimulate the various cones of the retina. Nor are there any sounds in nature; instead, variations in air pressure cause small filaments in the cochlea of the inner ear to vibrate, triggering nerve impulses. Both colors and sounds are illusions produced in the brain.

In the eighteenth century, the philosopher George Berkeley, while exploring whether objects could exist without being perceived, inspired the following question: "If a tree falls in the forest and no one is around to hear it, does it make a sound?"

The answer is no, because sound is a concept that the brain creates when it receives pressure variations from the outside world. But it is not just colors and sounds; the same goes for textures, tastes, and smells built up from the stimuli received by other sensory organs.

What we finally experience as a conscious visual scene, then, is a highly processed image very different from the informational flood received in the retinas of our eyes. That highly distorted set of pixels composed blurry, shifting images, initially captured in two dimensions and on a photoreceptor surface with a blind spot that we unconsciously fill in.

The brain is in charge of correcting everything: it elaborates a clear three-dimensional image without a blind spot, stable in relation to eye and head movements, and filled in and reinterpreted on the basis of our previous experiences of similar scenes.

We are never, ever, aware of all these operations.

Classical philosophers such as Aristotle, Plato, Ptolemy, the eleventh-century Arab scientist Alhazen, and, later, Kant were already arguing that we see only manifestations of reality, because reality itself is beyond our perception.[1] At the end of the nineteenth century, the scientist and inventor Hermann von Helmholtz deduced, from the observation of visual illusions, that the poor visual information we receive is elaborated in the more complex, unconscious inferences that dominate our perception. Thus, these illusions—and indeed, perception in general—cannot be consciously rationalized.[2]

Recent research has shown that Helmholtz's concept is not confined to the field of visual perception. This model describes a much more general form of cognitive processing that occurs not only in our other senses but when we are remembering or making decisions.

The Creative Processes of Our Brains: Feeling, Attending, Perceiving

Capturing details, extracting the essence of external reality or concepts, internal processing through association with our memories to create relationships and categories—these are all elements of a general mechanism that prepares the brain to think, reason, and make decisions.[3] As we explain later, these are "creative" processes: believed to be retrospective, they are based on previous information stored in memory and they are inferential, meaning that our perception is based on the elaboration of hypotheses beyond the limits of our sensory input. In this type of processing, language, a capacity unique to human beings, plays a strategic role: it allows us not only to communicate but also to create abstractions and categorizations.

As we have said, most of these processes are automatic and unconscious, taking place in a complex interaction between the detection of sensory stimuli and the subsequent processes of attention and perception, which we describe later in detail.

Let us explore in brief how we perceive and attend to information. We capture external stimuli thanks to a type of memory called sensory memory. Sensory memory briefly retains sensations, those first impressions of reality that invade the sense organs in the form of lights, colors, smells, and sounds. These sensations are fleeting, and only a small fraction of what we capture will be processed later. For instance, you may go to a restaurant and after a while be unable to remember which waiter took your order.

The brain must select from all the information it constantly receives, and it does so through a process of filtering and concentration that we call "attention." To focus our attention, the brain requires the use of short-term memory, which has a limited capacity. When we focus our attention, the brain is able to reach the next stage: perceiving.

Perceiving is a process that gives meaning, usefulness, value, and emotional weight to the sensations we are attending to. Meaning is obtained by integrating, making coherent, categorizing, and recognizing these sensations based on probabilistic inferences—that is, by deriving conjectures from our long-term memory.

The extraordinary thing about all the cognitive processes required to interpret reality is that they manage to be very accurate using little information: "I am not interested in stopping at the specific shapes of thousands of black hairs in contrast to as many yellow ones, I am interested in knowing that it is a tiger and running away."[4]

Perceiving, in short, is a creative process.

Of the seamless flow of information that we receive through the senses, only a small part, usually that which we attend to for some reason, happens to be perceived. For instance, when we enter a restaurant, we have at our disposal a huge amount of information; however, we will invariably look for the maître d' or a waiter to direct us to a table. Our attention will mainly focus on the people who, according to our prior knowledge, could belong to that category, and we will consciously per-

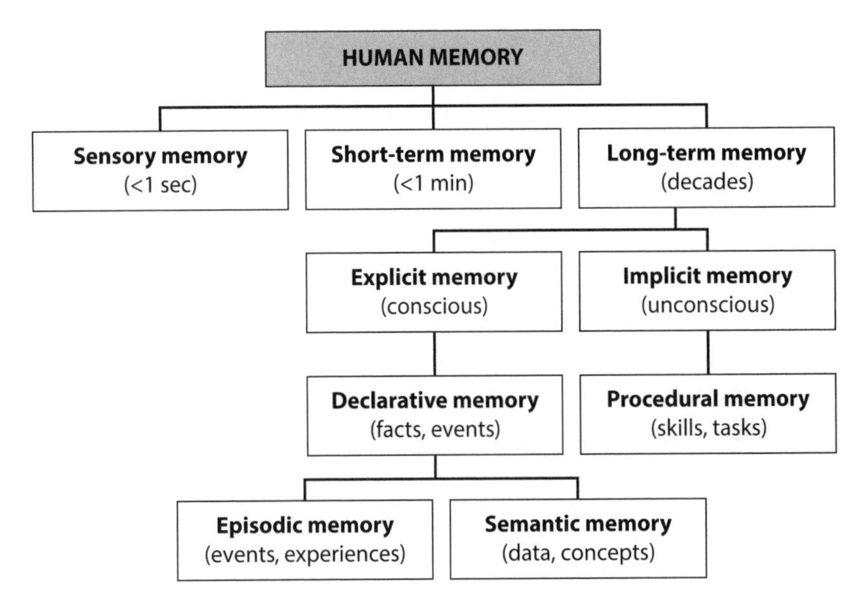

FIGURE 3.1. Types of memories.

ceive only such people, discarding practically everything else. This is an illustration of the sensation-attention-perception path by which conscious thoughts are elaborated. At each step, different types of memories must act in a coordinated manner.

Different types of memory, therefore, serve not only to save recollections and store learning but also to aid in perceiving and creating reality. In addition, they give our existence a sense of continuity. Only when the brain draws on its memories are we able to imagine or anticipate the future; this is why everything we feel and think is sometimes said to be the sum of our memories (figure 3.1).

Memories thus define our identity.

How the Brain's Memories Work

Memories are not only tools for archiving recollections and learning. They are also indispensable for speaking, thinking, and understanding the present; for imagining futures, planning, and projecting ourselves; and for making decisions. For this reason, we refer to "memories" in the plural, as they all express the existence of different brain mechanisms for managing information.

Memories are based on neural network circuits distributed throughout the brain. A concrete and stable memory corresponds to the activation of a group of neurons that are interconnected and synchronized with each other—what was once called an engram.

Depending on the connections, the same neuron can participate in different memory networks. As discussed in chapter 2, we now know that synchronization and connection are closely related concepts and that the learning process strengthens the connections between neurons, leading to structural and functional changes in the neural networks that form the basic units of cognitive information. In turn, these changes are activated and modified during the process of remembering.

As we have noted, some 85 billion neurons in the brain relate to each other through billions of connections. This synaptic plasticity truly defines the richness and versatility of the memory processes.

Let's explore this concept in more detail. Some of the connections between neurons are strong or durable, while others are weak. This versatility is what explains the organizational differences in different types of memories. For instance, in short-term memories, the strength and duration of the connections are weak or transitory, but in long-term memory, new proteins are synthesized, allowing networks with lasting connections to form.

If the neurons in a memory network are frequently stimulated, their synaptic connections increase in number and get stronger. This is called "long-term potentiation," and it underscores the importance of repetition and learning as the basis for the formation and storage of long-term memories.

In contrast, if memories are abandoned and not evoked, the connections weaken over time and what has been memorized fades, as we will see in chapter 8.

Given all of these concepts, how many types of memory could we say exist?

Following the well-known Atkinson-Shiffrin memory model, information management processes are structured in three stages or large types of memory: sensory memory, short-term memory, and long-term memory.[5]

For practical reasons, we will follow this classification scheme. However, we have no qualms about admitting that reality is much more complex and that the different mechanisms or types of memory overlap and act in concert.

Let's look at these types of memory in more detail.

Sensory Memory

Sensory memories are unconscious and ephemeral memories that retain for a very short time (tenths of a second) the information that the brain receives through sight, hearing, smell, taste, or touch. These types of memories record information immediately and are capable of discarding a good part of what is recorded so that they store only what makes sense to us.

We have different types of sensory memory, arising from our different senses. The most relevant are iconic, or visual, memory, which lasts a fraction of a second; echoic, or auditory, memory, which lasts for seconds; and haptic, or tactile, memory, which seems to decay after about two seconds.

Thanks to sensory memories, the so-called illusion of continuity is achieved—a strategy of the brain that allows us to have an integrated vision of reality. In fact, through iconic (visual) memory, coherent and quite faithful images can be formed; we merge and sequentially fill in the impressions obtained by our eyes while scanning a scene in a continuous movement. We explain this very interesting and little-known phenomenon more extensively in chapter 4.

Sensory information, after passing through the thalamus, goes to the different sensory cortices, where it is evaluated and processed before finally arriving at the prefrontal cortex. There it will be made available to the next type of memory: short-term memory.

Sensory information that is not attended to in short-term memory will be lost.

Short-Term Memory

Short-term memory allows us to retain a limited amount of information for a short period. It is transitory memory, and as magicians have learned very well, it is fragile and vulnerable to interference of any kind.

Short-term memory is also what generates the "flow of consciousness"—our perception of the present, or our concept of "now." We can simultaneously retain about seven different things or concepts, but not much more.

In other words, the majority of things or concepts retained in short-term memory are forgotten. Only a few are fixed, or consolidated, as long-term memories.

How long can this fleeting type of memory last? Estimates range from a few seconds to several minutes, the length of time depending on the meaning and quality of the information as well as on the motivation to remember it.

Different short-term memory systems are specialized in order to process distinct sensory systems and to facilitate language comprehension and various types of reasoning. A very important form of short-term memory, so-called working memory, allows us to retain and, in turn, temporarily operate or manipulate information. Consider how you are able to remember the content of a sentence while writing it, how you can mentally perform an arithmetic calculation, and, in general, how you think and reflect.[6]

Working memory is one of the key structures and systems of the brain that allow us to organize ourselves, turn instructions into plans, make decisions, and reason in general. This is why one of its functions is to focus our attention in the midst of distractions. Its centrality is also probably why it has been found that working memory overload can hamper creativity and limit the capacity to find relationships between apparently disparate elements.

Short-term memory does more than simply process information in the present moment. Information from long-term memory—that is, from our knowledge and past experiences—can only be accessed through the short-term memory buffer. For this reason, working memory operates not only in the present but also in the future, anticipating objectives, plans, and events.

Furthermore, working memory is capable of resolving present situations by making inferences based on past experiences. This broad working portfolio explains why there can be individual differences in the capacity for working memory; some studies have even correlated it with measures of intelligence.[7]

Long-Term Memory

Long-term memories are those in which we store a large amount of information over a long period of time—minutes, hours, or even indefinitely. They are more stable and durable and less vulnerable to interference than short-term memories, but some long-term memories can still be manipulated in so-called reconsolidation processes.

Many long-term memories are obtained gradually as information stored in short-term memories is repeated, forming neural networks with lasting connections.

There are two main types of long-term memory, explicit and implicit. Explicit or declarative long-term memories are of a conscious nature and include semantic memories (of concepts, names) and episodic memories (of events, occurrences, experiences). Implicit long-term memories are unconscious: they include memories of procedures or motor tasks, such as learning to swim, as well as perceptual or emotional memories.

These two types of long-term memory, explicit and implicit, work simultaneously. We are beginning to understand the formation (and evocation) of memories and the generation of forgetfulness—processes that, by the way, are not overlooked in some magic effects (as we will see in chapters 8 and 9).

Emotions

Emotions are global, automatic reactions of the human organism. They are generally intense and of an affective nature, provoked by external factors or internal thoughts, and they often manifest through a visible, organic jolt. We can distinguish between primary and secondary emotions.

There are some six primary, or basic, emotions: joy, sadness, fear, anger, disgust, and surprise. All are essential emotions that, from an evolutionary point of view, are critical to our survival, well-being, and effective decision-making—in other words, for triggering a “fight-or-flight” response and for motivating our search for food, sex, social bonds, and so on. Emotions are also innate—biologically determined—and as Darwin observed in his studies, each of them has a characteristic facial expression common to all humans regardless of their ethnicity or culture.

In this sense, primary emotions constitute an involuntary system of signs for expressing ourselves to fellow human beings as well as to other species. Their expressiveness makes emotions the first means of nonverbal communication between individuals.

Besides the six primary emotions (a list that generates quite a bit of consensus), there are other types of emotions that are considered secondary because they arise socially through cultural conventions. These emotions, which generally last much longer than the primary ones, include guilt, embarrassment, shame, pride, envy, jealousy, admiration, and compassion.

Primary or secondary, every emotion has its own physiological or characteristic stamp. Their objective components can be of two types, vegetative and motor. The vegetative component is activated by the autonomic and endocrine nervous system, which causes sweating, vasoconstriction, and pallor. The motor component is characterized by facial or vocal expressions, gestures, or postures. Emotions also have behavioral components, such as impulse, or the inclination to act.

Feelings

All emotions are accompanied by feelings, reactions that constitute the subjective awareness of emotion.

Feelings are more complex and elaborate reactions than emotions. They are the second moment—think of the indignation that comes after anger. The neuroscientist Antonio Damasio considers it important to make an express distinction between emotions and feelings, lest we confuse an emotion with the related feelings that often follow.[8]

When we can sense the type of emotion we are experiencing, we can better regulate our emotional response—our feelings—and that allows us to shape our affective behavior. In other words, emotions cannot be controlled, but the same is not true of the feelings associated with them.

Despite how little control we have over emotions, they are determining factors in our motivations for future behavior, whether it is healthy behavior, such as altruism, or pathological behavior, such as obsession or addiction.

Emotional Memories

Emotions are processed in the amygdala in the form of implicit memories. These are the so-called emotional memories, which are discussed in more detail in chapter 9.

Unlike explicit or declarative memories, emotional memories are unconscious, faithful, rigid, and lasting. They preserve innate emotions, such as fear of snakes, and inherited conditioning, such as disgust over rotten food, as well as acquired habits, such as tobacco addiction or an obsession with cinnamon.[9]

The interrelationship between emotional and cognitive mechanisms, between "passion" and "reason," is extensive and explains why rationales, decisions, and many other cognitive functions cannot be understood apart from emotional components and emotional memories.

We consider emotions here not only because they have a significant influence on attention, memory, learning, and decisions, but also because they are inherent reactions to the outcome of magical effects.

As we will see in the following chapters, emotions can make it difficult to pay attention and can cause us to become distracted. They may influence the fixation or consolidation of memories, but they can also influence the recall of memories. This makes it possible for emotions to alter, to a great extent, the mechanisms of reasoning and decision-making in situations of anxiety and stress, for example, or in reactions of anger.

In chapter 11, we explore the magic act itself and what audiences feel after the outcome of a magical effect. In a magic trick, the first emotion aroused after an impossible outcome is surprise, an emotion that has an innate defensive function—as in the face of danger, for example. Surprise allows one to interrupt the action in progress, to orient oneself toward the significant event, and then to react.

Surprise is the first emotion in the face of an unexpected event, be it positive or negative, or of low or high intensity. This emotion lasts for a very short time and is followed by other emotions (fear, joy, relief, or confusion), as well as a fight-or-flight response (as when faced with the sudden appearance of an angry lion). Surprise is also a brief, transitional emotion in response to a magic trick and is immediately followed by other reactions that can range from disbelief to confusion to laughter, joy, or uneasiness.

But surprise is not the only emotion in magic: the illusion of the impossible is an intellectual and emotional experience unto itself, and one that we explore at length in chapter 11.

For now, with these basic concepts of neuroscience as a base, let us go into the cognitive processes involved in magic effects.

PART II

The Mechanisms

CHAPTER 4

We Build an Illusion of Continuity

The Limits of the Brain and the Illusion of Continuity

The brain assumes that the world in which we live is continuous in space and time. Our sensors, however, do not allow us to experience it that way. As we have discussed, the limits of space and metabolic consumption prevent us from simultaneously processing all aspects of a scene in high resolution.

In addition to the brain's restrictions on processing information, the visual field itself has limitations that force us to constantly move our eyes, even if we are not aware of doing so, to fill the continuous, inevitable gaps in our vision. This eye movement is what gives rise to the so-called illusion of continuity, as understood in both space and time.

Let's take an example: when we see a car moving down the road, even if we perceive it completely, we actually process only a part of the vehicle in high resolution. The same happens with the movement of the car: we experience an illusion of continuity in time because the brain, lacking the ability to process all the information that constantly comes in from the outside world, processes mainly contrast—that which changes or is in motion.

As we take in the scene, we perform various imperceptible scanning movements with our eyes that manage to fill in a good part of the visual experience—the rest of the car, the road—on their own. This efficient strategy achieves great results while consuming little energy and taking up little space on our "hard drive." With very little information, we experience the illusion of a car flowing smoothly on a road.

The Particularities of the Field of Vision

Our field of vision—the portion of space that each eye is capable of seeing—is very wide. Owing to the unequal distribution of different cell types in the retina, however, it is also asymmetrical. And as we have discussed, our brains "fill in" in the blind spot, where the optic nerve begins.

In reality, we have good visual acuity only in the fovea, a central region of the retina with a very high density of photoreceptors (cones) that represent only 2 percent of the total retinal surface. In the rest of the retina, the acuity is so low that, if we lacked the fovea, we would be declared legally blind.

This uneven distribution of visual acuity makes our peripheral vision much less clear and lacking in detail. For instance, when you look at someone's face, your central vision is fixed only on their eyes or some other specific aspect, such as their nose or mouth, while the rest of the face remains a low-resolution cloud.

A clear example of the difference between central vision and peripheral vision is offered by Margaret Livingstone in her book *Vision and Art: The Biology of Seeing.*[1] This work is a study of Leonardo da Vinci's most famous painting, *La Gioconda,* also known as the *Mona Lisa.* When we focus on the eyes of Mona Lisa in this painting, her mouth is in the periphery of our vision, where it is much less precise. We therefore perceive the shadows that surround her mouth as enhancing its curvature, creating the illusion of a subtle smile. By contrast, when we look directly at her mouth, we are able to differentiate the shadows from the corner of the lips, and her smile fades.

We are limited by more than the concentration of photoreceptors in the fovea. Retinal neurons detect and process information primarily when there are changes in the scene; when the scene is static, the receptor cells adapt and switch off. This happens because the retinal pigments of the photoreceptors become exhausted when they absorb light and need a moment of darkness to regenerate.

The only way to "see" static images and analyze a scene in its entirety, then, is to make continuous eye movements, thus letting the fovea scan different parts of the scene sequentially and avoiding the consequences of adaptation.

These "scanning movements," in turn, are closely related to the information that we are interested in obtaining at a given time. The scientist Alfred Yarbus made this clear in his book *Eye Movements and Vision.*[2] He observed that we tend to focus on an image's salient aspects, those that contrast, and that the information processed in fixation depends on the tasks or instructions received. In other words, looking for something in a picture is different from memorizing its content.

Yarbus illustrated his theses through an experiment with the painting *Unexpected Visitors* by Ilya Repin. His experiment consisted of tracking spectators' eyes after asking them specific questions. The results revealed that, for example, when the spectators were asked about the age of the characters in the painting, their eyes were directed at the characters' faces, but when asked about the characters' economic situation, the spectators looked at their clothing or material possessions.

Another classic experiment, proposed in a 1978 essay by Richard Anderson and James Pichert, demonstrated that the reconstruction of a scene depends very much on the objective of the search—that is, on the instructions received.[3] The participants in their experiment watched a video showing a tour inside a residential mansion. Some viewers were instructed to watch the video with the eyes of a thief, and others were told to watch with the eyes of a potential buyer.

After the viewing, the memories of the two groups varied considerably: the potential "buyers" looked at the number and size of the rooms, their facilities, bathrooms, and the like. Potential "thieves," on the other hand, looked at the accessibility of windows from the outside and at potentially sellable objects, such as television sets or stereo systems (these were earlier times). The participants' memory of the video tour was imperfect in both groups, differing according to the task at hand.

The Various Types of Scanning Movements

Among the various types of eye movements, the following three stand out:

1. Saccades, or quick movements
2. Smooth pursuit movements, made when tracking a moving object
3. Micro-saccades, the small and involuntary movements made during fixations, when the eyes' gaze temporarily stops in a certain position (between 180 and 330 milliseconds) that keeps the receptor cells from adapting and switching off[4]

Of course, we do not perceive the movements of our eyes—if we did, we would be afraid of ourselves. Importantly, most processable information is obtained during the movements of fixation. From partial, fragmented images, the brain must fill in the missing information to provide the illusion of continuity in both space—given that we never scan a scene in its entirety—and time. The temporal filling in is especially relevant since we are basically blind to the world between fixations. That is not to say that we are completely blind to the world; this is not an all-or-none situation. But it is true that the information we gather during saccades is poor compared to what we obtain during fixations, and that we mostly use information gained during saccades to distinguish what part of our perception is the result of the movement of our eyes and what is the result of genuine change in the scene during the fixation.[5]

Let's do some basic calculations. If we perform an average of 3 or 4 micro-saccades, or saccades per second, and we are awake some sixteen hours, we will spend four of those hours visually disconnected because rapid eye movements last between 150 and 300 milliseconds, depending on their amplitude. To this we must add an additional half-hour of darkness due to blinking, which we do every five seconds, with an average duration of 500 milliseconds.

Thus, although we see the world in a discontinuous way, our experience is not fragmented but continuous in both space and time. The question that follows is this: How do our brains merge the different visions of a scene?

The Image Fusion Process

It is memories that allow us to fuse different visions into an image. First, memories allow us to properly format the sensory information we receive and make it available for a few moments—tenths of a second—before it disappears completely. And second, memories allow us to fill in the gaps between successive images on the basis of information already stored about the world around us.[6]

The process of merging and filling in is so efficient that we can easily calculate the speed and direction of a moving object, making it possible for us to see reality as an unbroken sequence rather than as a succession of independent, static scenes.

The invention of the thaumatrope, a toy that was very popular in Victorian England and first described by John Ayrton Paris in 1825, illustrates the fusion of successive images. A thaumatrope is a flat piece, usually made of cardboard, wood, or metal, with two different images, one on each side (for example, a bird and a cage, or a bouquet of flowers and a vase). By quickly rotating the piece around on its axis, the two images are integrated into one.

This fusion of successive images is probably also the principle behind the Spanish magician Dani DaOrtiz's version of a trick by the nineteenth-century Peruvian magician known as "L'Homme Masqué."[7] In this trick, the magician rifles before the spectator a deck of cards containing the entire suit of hearts in order, except that the queen of hearts has been replaced by the queen of diamonds. This very subtle change of card is so imperceptible that the spectator thinks that they have seen the queen of hearts instead of the queen of diamonds.

Magic also takes advantage of the existence of sensory memories in other sense organs, such as the sense of touch. For example, consider a magician removing a watch from a spectator without the spectator realizing what is happening. Before unbuckling the strap, the magician presses the watch face against the spectator's wrist, thus generating a "post-sensation," which prevents the victim from realizing after the removal that the piece has been stolen because for a few seconds it feels as if they are still wearing the watch.*

These sensory memories affect the process of "contrast gain control," a term used in neuroscience to explain how the perceived intensity of a stimulus is influenced by what has occurred immediately around it, in both space and time.

The Illusion of Continuity and Cinema

The fleeting nature of iconic memory could explain the "phi phenomenon": an optical illusion that allows us to perceive continuous movement where what really exists is a mere succession of static images.

What is the basis for this phenomenon? If we are shown fewer than twelve images per second, we perceive them as separate images, like the blink of an eye, but increasing the frequency generates an illusion of continuity in time that makes the succession of images seem to be moving. In analog cinema, for example, all movement is inferred from a rapid succession of different frames or static images.

In analog cinema's most conventional form, we see twenty-four different images per second, and the blinking sensation is avoided even more effectively because we are shown each image twice—saving half the film!—separated by dark frames of the same length. More than one hundred years ago, Thomas A. Edison, a pioneer in the reproduction of moving images, proposed that using 46 pulses of light per second would avoid blinking. At this frequency, movement appears natural because each static image, each frame, cannot be consciously perceived in isolation.

In silent films, the frame rate was manually controlled and therefore was variable, sometimes even intentionally variable. The final speed was controlled by the projectionists, who, if they had a "good eye," could synchronize the speed of projection with the speed of capture. It was not uncommon, however, for projectionists to speed up the frame rate

* Post-sensations are systematic changes in the perception of a stimulus after adaptation to the presentation of a previous one.

a little to save a few minutes and gain one more showing per day. A variable frame rate could also serve more artistic purposes. The speed was often varied in different parts of the same film to make action sequences seem faster and more frenetic, and romantic sequences more leisurely and contemplative.[8] The variability in the speed of projection disappeared with the arrival of sound: because the soundtrack was integrated into the projection tape, the dialogue and music had to sync naturally with the images.

Today we continue to seek an improved visual experience by manipulating the frequency of frames played per second. Movies are being marketed at over 48 frames per second, and experimental technology exists that allows recording at extraordinary speeds, some 100 to 120 frames per second, in what is such a rapid succession of images that even the unconscious perception of change is impossible.

In short, the illusion of continuity allows us to enjoy the movies and, as we will see later, to enjoy many magic tricks.

The Illusion of Continuity and Sound

Humans hear sounds in the range of 20 to 20,000 hertz. The lowest note on a normal piano, for example, vibrates at a frequency of about 27.5 hertz, which is about the minimum frequency necessary for us to see a continuity of moving images. (Analog cinema is a succession of about 48 frames per second.)

When the molecules vibrate around that speed, we hear something that sounds like a continuous note, demonstrating that we also build an illusion of continuity with sound by unconsciously filling in gaps in the auditory data. To quote Daniel Levitin from his book *This Is Your Brain on Music*:

> If you put playing cards in the spokes of your bicycle wheel when you were a kid, you demonstrated to yourself a related principle: At slow speeds, you simply hear the click-click-click of the card hitting the spokes. But above a certain speed, the clicks run together and create a buzz, a tone you can actually hum along with; a pitch.[9]

The Illusion of Continuity: A More General Process

Extracting, interpreting, filling in, and constructing an illusion of continuity not only occurs across the senses but is also a general principle of brain functioning. For example, if you are in a busy place, your ears receive a cacophony of voices in the form of superimposed sound waves, yet, to your surprise, your brain has a remarkable ability to identify them.

The brain can do this because, as with visual stimuli, it extracts certain features and integrates them into its own interpretation of what it is hearing. If part of a word in a sentence is blocked—for example, by a cough—the brain engages in "phonemic reconstruction." Similar to the filling in that the brain does when it masks the blind spot of the retina or when it improves the low resolution of peripheral vision, phonemic reconstruction "fills in" the missing sound.

Phonemic reconstruction, like visual restoration, depends on the context in which we hear the words, so the word you think you heard at the beginning of a sentence may condition the word you hear (masked) at the end. Taking an example from everyday life, imagine hearing the phrase "the hunter saw the hog"; you will surely have no problem understanding that the hunter visually detected the hog. But if the phrase starts "the carpenter . . . ," you will tend to interpret the same sound, "saw," as "sawed."

As detailed in chapter 8, similar brain processes are at work when we evoke memories, unconsciously and automatically filling in details. For example, when you see a person for the first time, you may imagine what their personality is based on their appearance, their ethnic group, or the fact that they look familiar.

Change Blindness

The illusion of visual continuity also involves an interesting phenomenon called "change blindness."

When we visually reconstruct a scene by continuously moving our eyes, we introduce artificial changes in the information that each visual neuron receives. Visual neurons are designed to detect changes, and

FIGURE 4.1. "Gradual Change Test."

they remain inactive when visual stimuli stay constant, a phenomenon known as adaptation. The visual stimuli on the retina change not only when something changes outside, but also when we move our eyes, even though the external images remain stable. Our neural circuits are designed to discount these eye movements and to act as if they were momentarily blind to the effects of those changes, whether they happen very slowly or coincide with some kind of visual discontinuity.

Scientific studies on change blindness have shown that large changes in the visual scene can go unnoticed by us if they coincide with a transitory interruption, such as a blink, an eye movement, flashes in the scene, or sudden changes in the direction of movement, even when we are looking in the right place.[10] Very gradual and uninterrupted changes can also go unnoticed, as shown in the video "Gradual Change Test" (figure 4.1).

In this illusion, and without our realizing it, the floor of the carousel changes very slowly from a reddish hue at the beginning to a more purple hue at the end of the video. We experience this "blindness" because of the inherent limitations of short-term memory. Short-term memory cannot process all the information it receives, and it cannot compare, in real time, past information with present information—here, the new scene with the immediately previous one—unless the changes are significant and capture our attention.[11]

But in fact, there is usually little need for our short-term memory to function in this way. The world does not normally change substantially while we make rapid eye movements or while we blink, so we have not acquired a brain circuit specially designed to detect such changes because they are largely irrelevant to survival. Evolution has chosen to

FIGURE 4.2.1. "Whodunnit?"

FIGURE 4.2.2. "The Color Changing Card Trick."

dedicate our scarce brain resources to better purposes. In other words, although it produces some side effects, being fundamentally blind to a large part of the changes taking place around us is advantageous in terms of evolution.

Change blindness, in fact, is the reason why some specific real-life situations can be dangerous. Take, for instance, bicycles in urban traffic, which can go dangerously unnoticed by drivers. To reinforce the safety of cyclists, Transport for London posted an informational video in 2008 using change blindness. Very much in keeping with British sensibility, the video, entitled "Whodunnit?", consists of a theatrical presentation of a crime in which changes occur at each camera jump, many of them undetectable by people seeing the spot for the first time. (see figure 4.2.1).

A video posted in 2012 by the British magician and psychologist Richard Wiseman, "The Color Changing Card Trick," is still circulating on the internet. As Wiseman himself states at the end of the video, he is not performing a card trick but rather demonstrating change blindness (see figure 4.2.2).

John Henderson and Tim Smith, researchers at the University of Edinburgh, analyzed where viewers fixed their gaze while watching Wiseman's video by using an eye-tracker device (see "Color Changing Card

FIGURE 4.2.3. "Color Changing Card Trick and Eye Tracking."

FIGURE 4.2.4. "Princess Card Trick."

Trick and Eye Tracking," figure 4.2.3). Those who saw the video for the first time and did not notice the changes looked at the same places as those who had already seen it. Interestingly, in a study with a different magic trick, the opposite result was observed: eye movements during the critical moment of a coin trick showed a different pattern depending on whether the participants had previously seen the secret or not. Either way, both studies suggest that laypeople are blind to changes in a scene when they first see it.[12]

Magicians take advantage of change blindness for their own purposes. For example, they can count cards in so ingenious a way that the audience does not notice that a particular card is repeated during the count. Henry Hardin's classic "Princess Card Trick" is a good example of how magic takes advantage of change blindness (see Lance Burton's version at figure 4.2.4). In this trick, the limitation of our short-term memory prevents us from comparing the first sample of cards with the second sample, in which none of the cards from the first sample appear. The trick relies on this dramatic change going completely unnoticed. Indeed, the risk that spectators will detect this type of change is lower when they are actively involved in the game.[13]

Prestidigitation: Is the Hand Faster than the Eye?

Centuries ago, magicians empirically discovered that by manipulating the speed of their maneuvers they could make them invisible. As the saying goes, "the hand is faster than the eye." Yet again, magicians did not need to know neuroscience; by trial and error they went straight to the results.

Indeed, magicians are capable of many fast maneuvers that cannot be detected visually as they sneak or conceal cards, coins, balls, or other gadgets from the eyes of spectators. Such manipulation probably gave rise to the word "prestidigitator"—a sleight-of-hand magician, from the French *preste* ("nimble") and the Latin *digitus* ("finger"). For example, magicians executed coin magic at the end of the twentieth century with very fast maneuvers, such as catapulting coins from one hand to another. Only after the introduction of the new types of trick coins was "slow" coin magic possible.[14]

The effect of rapid movement in magic was first studied scientifically in France in 1893, thanks to the support of the newly founded French Association of Prestidigitators, presided over by the magician and theater and film director Georges Méliès. Méliès has gone down in history for his innovations in the early days of cinema, such as special effects like multiple exposures, substitution splices, time lapses, and dissolves, and also for his narrative developments and surrealist fiction films inspired by the travels of Jules Verne.

Méliès bought and directed the Paris theater of the famous magician Jean Eugène Robert-Houdin. He made it possible for two renowned magicians of that time, Arnould, a great mnemonic professional, and Raynaly, a stage magician, to collaborate in the experiments of Alfred Binet, a psychologist. Binet was famous for introducing, among other advances in psychology, the Binet-Simon IQ test, the first test to evaluate intelligence.

These were times of tension between magicians and mediums. In seeking to discredit the mediums, who were presumed to possess supernatural powers, magicians began to support scientists interested in the mechanisms of magic. Thus, in 1894, Binet published some pioneering results after applying chronophotography to various secret techniques of the magicians of that time.[15]

Chronophotography, an emerging technique of time-lapse photography, was carried out with very rudimentary instruments that were essentially cameras moving a short strip of film over a primitive lens using clockwork mechanisms. Binet studied "sleight of hand" and other maneuvers with a system that allowed him to take between ten and fifteen photographs per second. Thanks to this method, he could calculate that Arnould had made a very quick pass (to bring the desired card to the top of the deck) and also announce that he had done it in a tenth of a second!

Most important, however, Binet discovered that time-lapse photography diminished the effects of magic, because the effectiveness of many tricks depended on the inability of the eye to perceive rapid changes. His published results showed that magic's illusion lies not only in a movement's precision but also in its speed.[16]

Slow Magic

Nowadays, however, slowness is increasingly prestigious in the world of magic. It is now well known that magicians do things rapidly to fool the audience, so a slow trick makes the magic act more convincing. The magician Ascanio always said that magic tricks should be performed not only with clarity and ease but also in slow motion, although not so slow as to be boring: "The good swimmer is the one who glides smoothly through the water without splashing." Slow motion, he said, "gives credibility, yet makes the public uneasy, believing it is in possession of all the facts."[17]

Another magician who made his maneuvers in slow motion was the Argentinean René Lavand, who elevated slowness as a magician strategy with his "lentidigitation"—sleight of hand executed in slow motion, from the Spanish *lento* ("slow"), replacing *presti* in "prestidigitation"—and whose famous tricks included one called "You can't get any slower." In a related study with small paddles and popsicle sticks, decorated with marks and signs that magicians can make appear and disappear, it has been shown that, when we are faced with two simultaneous trajectories of movement, downward movement and rotation (of the paddle), and the turn is executed at a speed close to the resolution limit (that is, close to 60 hertz), we stop seeing the rotation and perceive only the downward movement—that is, the slower movement.[18]

In summary, visual continuity, a concern for magicians, can be affected by change blindness (owing to the limitations of short-term memory), as well as by the invisibility of objects caused by fast movement. Magicians challenge the two limits of vision with respect to their speed: on the one hand, they make very fast manipulations that cannot be processed by the brain, and on the other, they use slowness so exaggerated that, as we have seen, it can generate blindness to change.

As we will see in chapter 6, the movements of objects in a scene can be effective attention grabbers, and when several movements overlap in a single scene, whether or not they are perceived will depend on their speed and timing.

CHAPTER 5

Magic and Contrast

The Key to It All

The Funny Thing about Magic

Structurally, a magic effect is similar to a joke. Like magic, a joke is a step-by-step narrative, a series of logical and predictable events that concludes with an unexpected twist. The resulting absurdity provokes laughter and requires a complete reinterpretation of what came before.[1] The humor in the telling of a joke depends on how the story ends. When you hear it for the first time, you cannot complete the story on your own or anticipate the ending. The same is true of magic: the viewer does not know what the final effect will be, so the end is usually a surprise.

The impossible ending of a magic trick dazzles you, just as a joke's absurd conclusion makes you burst out laughing. Crucial to both are rhythm and timing. If you tell a joke too fast, the brain does not have time to imagine a possible outcome, which does the important work of providing a contrast with the unexpected ending. On the other hand, if you tell the joke too slowly, the brain has time to imagine various scenarios, and the joke may lose its final punch. The bottom line is that, with both magic and humor, the audience's ability to imagine the future undergoes a sudden and surprising disruption.

Contrast and the External Life of a Magic Effect

Any magic trick or effect has a narrative structure and a resolution or climax; in other words, it is a kind of expository presentation that ends with an impossible outcome. A magic trick can have more than one effect, however, so routines typically go through a set of consecutive effects or tricks.

The type of climax at the end of the effect of interest to us here is the illusion of impossibility. Our focus is thus on the magic effect that has an impossible outcome and lasts for some seconds or even minutes.

Magic tricks have a double reality. Following the ideas of the magician Arturo de Ascanio, we can distinguish between the "external life" of the effect, which would be what the audience sees, and its "internal life," that is, the secret actions that make the whole effect possible.[2]

The effect's external life includes the presentation or expository phase during which expectations for the climax are created. This expository phase may tell a story, or it may be merely a demonstration. In many effects, the expository phase is highly plausible because what is presented to the viewer is logical, with predictable consequences.

What really happens during the expository phase, however, is that the magician, as he guides viewers to a surprising outcome, is hiding his deception by using materials and methods that the audience cannot see. In short, this is what constitutes the internal life—a secret, parallel reality that is not seen and that involves concealments, maneuvers, and other procedures, sometimes with the help of gimmicks and gadgets. These materials and methods are part of an extensive catalog, carefully guarded for centuries by magicians, that is unique to the world of magic.

The first condition of a good magic effect is that it be properly structured and that, in its external life, everything that happens is logical. It is just as important that the magician handle the techniques and gimmicks with dexterity and that the presentation of the effect be impeccable from all points of view. Ascanio emphasized that the magic effect consists of "the contrast between an initial situation and a final situation," but for this contrast to happen, contrast situations must be minimized throughout the expository phase so that obvious discontinuities do not distract the audience at key moments.[3]

To minimize this contrast, everything—the content of the expository phase, the structure of the effect, the character, the script, the presentation, and the techniques, materials, and specific methods used—must be selected and coordinated very precisely. When we refer to "contrast," however, we mean exclusively the contrast during the expository phase of the external life, not what Ascanio defined as the difference between the beginning and the end of the magic effect, or the "global contrast."[4]

Let us explore in detail the concept of contrast as we understand it, and why it is the determining factor in the neuroscience of magic.

We See Relatively, Not Absolutely

> I don't paint things. I only paint the differences between things.
>
> —Attributed to Henri Matisse

Brain networks are characterized by the presence of physical bottlenecks—that is, regions in which the information elaborated and processed by many neurons is transmitted much more sparsely by a significantly smaller number of cells, to later be amplified again in the subsequent stages of processing, saving physical space and metabolic cost. To manage visual information, the visual system of the brain encounters a series of such physical bottlenecks, beginning when information received in the retina from the outside passes to the brain through the very narrow optic nerve.

Bottlenecks require the filtering out and discarding of a lot of information, a process that generates successive phases of compression and decompression as the image travels through the brain. Inevitably, compression of the acquired image results in a loss of resolution, which can only be recovered by a decompression process called interpolation. Interpolating consists of generating new information within the filtered image to restore its resolution, but this process results in another side effect: loss of sharpness.

Let's explore how and why this happens by returning to the problem of space in the optic nerve. In chapter 2, we talked about the photoreceptors in the retina, through which humans capture visual information. You have about 100 million photoreceptors in each eye, arranged so that they cover the surface of the retina more or less evenly, as the pixel map does in the sensor of a digital camera. Photoreceptors capture photons and transform that physical stimulus into a local code of electrical signals that the brain can understand. The problem is that if you wanted to send individualized information from each of those 100 million photoreceptors, you would need 100 million "wires"—the axons of the retinal ganglion cells forming the optic nerve, which connects the eye to the brain. This would require an impossible amount of space and energy. An optic nerve with 100 million axons would have a diameter similar to that of the eye itself and would occupy approximately 40 percent of the available volume in the brain. In addition, the structures of the brain responsible for receiving and processing this information would have to increase their size accordingly, and you would end up with a head the size of a delivery van, which would be structurally unfeasible and require unsustainable amounts of energy.

A very important compression process takes place in the retina, the first physical and functional bottleneck: instead of sending 100 million axons to the brain from each eye, it sends only one million. Our "camera," after all, has only one megapixel! This massive compression process reduces the amount of space and energy required, but it also leads to very significant information losses, unlike sophisticated artificial compression algorithms. It so happens that, as information is transmitted to the brain, the information lost must be refurbished to reconstruct the image. We take high-resolution photos in the retina, transmit them at low resolution through the optic nerve, and reconstruct them in the cerebral cortex to their original resolution through interpolation. The missing information is filled in with, among other things, new information taken from the immediately neighboring pixels in the image.[5]

However, interpolation always involves a related problem: the final image gains resolution but loses sharpness. Increasing the size of the image and generating new points among the actual points necessarily

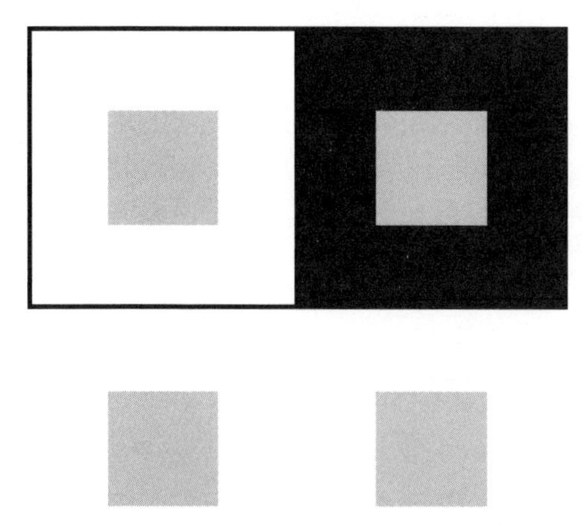

FIGURE 5.1. We see in a relative, not an absolute way: both gray boxes in the center reflect the same amount of light; however, we perceive it as lighter or darker depending on its context.

affects the quality of the image. For example, in transition zones between two extreme shades, such as white and black, interpolation creates gray-shaded pixels, which are an average of the different shade values of the white border and the black border (see figure 5.1).

This form of filling in by reducing the contrast between two real shades seems reasonable. In this way, you no longer go directly from white to black but from white to gray and from gray to black. This reduction of contrast leads, however, to a loss of sharpness, because if we look closely, the edges appear blurred. It is akin to "net diffusion," that old trick used in television and photography to make portraits of famous people with their wrinkles and skin defects reduced by stretching a piece of stocking in front of the camera lens. Our brains, to compensate for this negative consequence of interpolation, this loss of sharpness, compare the value of the image at each point with what is around it so that they process only the changes, the differences—or what we call contrast. Therefore, in practice, we see relatively, not absolutely, because the brain's circuits have become efficient contrast detectors. You can experience this effect by looking at figure 5.2.

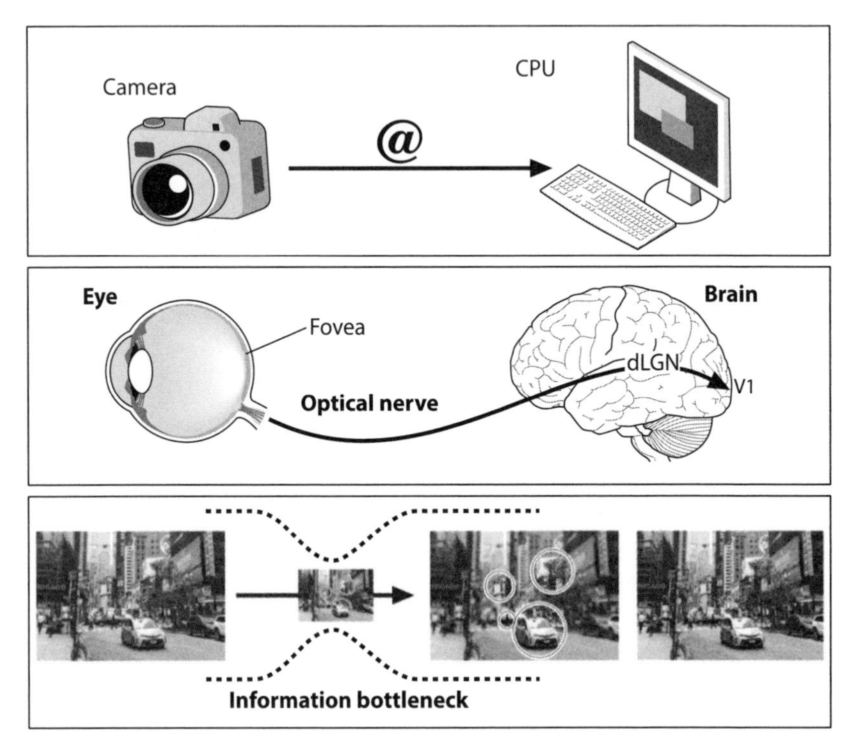

FIGURE 5.2. Interpolation and bottlenecks in a diagrammatic representation of the transmission of information along the initial section of the visual system: Retinal photoreceptors take a high-resolution raw image, which is then compressed before being sent through the optic nerve and reconstructed in the primary visual cortex (V1). This biological process is analogous to the transmission of information between different digital devices, such as a camera and a computer.

Contrast Detectors

So how do these contrast detectors work?

We have already noted that the retina prioritizes those points in space where the light changes in intensity. To do this, it uses two types of detector neurons called "on" and "off." "On" neurons detect local changes from less luminance to more luminance. Their preferred stimuli are those areas that go from dark to light. When the brightness is evenly distributed, these cells simply do not respond, and the retina sends no signal to the brain. "Off" neurons behave in a complementary manner, detecting local changes from more luminance to less luminance. The two "on" and "off" circuits inhibit each other: when one is active, the other is suppressed, further increasing the sense of contrast and the quality of the image and thus restoring its sharpness.[6]

Similarly, other areas of the brain have neurons that have specialized in detecting local changes in various aspects of an image, such as color, size, shape, and movement, and even attributes at a higher cognitive level, such as identification of objects.

Because bottlenecks are ubiquitous throughout the brain, and because "on" and "off" circuits are necessary to overcome their undesired side effects, there are similar "on" and "off" circuits controlling such disparate behaviors as the sensation of fear, decision-making, and motor behavior. Like retinal cells, these neurons in other brain networks do not respond to the absolute values of the parameters they process but rather are detectors of contrast and change, so they remain inactive when everything is constant in a visual scene.

A famous saying among neuroscientists who study the visual system is that a hungry toad surrounded by dead flies would starve to death; because the flies are motionless, it would not see them. With zero contrast, we would not see or pay attention to anything. Think of animals that can mimic their environment using camouflage, thus achieving the greatest reduction in contrast with the environment. Something similar occurs in black-light theater, where objects and people dissolve into the black background, becoming invisible, as if they did not exist. This is why, in a scene, the points of maximum contrast, the areas of maximum change, both spatial and temporal, are those that capture our attention—a fact that magic uses to its advantage.

Seeing in a relative and not absolute way has other undeniable adaptive values, such as enabling us to read the same text with no problem in very different lighting circumstances, such as in artificial light in a room or in full sun on a beach or terrace. In both situations, we have no problem perceiving black letters of minimum luminance (that hardly reflect light) on the brightest white of the printed page. And we are able to perceive the black letters despite the fact that the small amount of light reflected by them in the sun can be significantly greater than that reflected by the white page inside a house. Because a light bulb emits much less light than the sun, there is less light to be reflected inside the house. On the other hand, if we were to base our perception on absolute measurements and not on contrasts, black in the sun would seem brighter than white inside a home, and this would prevent us from distinguishing the difference between black and white, regardless of the situation we were in. In other words, we would not be able to generalize. Fortunately, this is not the case. To be consistent, our perception is based on a comparison between black and white according to each situation and at the local level. Because it has emerged as an adaptation to information bottlenecks, this ability to adapt to different visualization conditions, called visual constancy, is common to all types of brain processing.

The strategy of processing information in a relative but not an absolute way, of detecting contrast both in space and in time, has arisen to compensate for the negative effects of interpolation—the loss of sharpness associated with the process of compression and decompression on both sides of a bottleneck. These bottlenecks have arisen, in turn, as the most viable solution for sending large amounts of information over long distances at minimum cost, in terms of both space and energy consumption.

These energy-saving strategies are ubiquitous in the brain, even at the risk of losing information—including information that conditions not only how we see but also what we pay attention to, how we taste things, how we caress things, and even what we remember, as we will see in chapter 8.

Contrast Depends on Context

Contrast and context are inseparable, because any contrast is always between "the object," whatever it may be, and its "context" (see figure 5.1). The contrast may be spatial or temporal; it may also be explicit (present in the physical reality of the scene) or implicit (stored in our memories).

In fact, one of the strongest contrasts, cognitively speaking, and one that is most likely to get our attention, occurs when something in a scene breaks with our expectations, when things happen in a way that alters the logic we have experienced and learned in the past—as when magic creates the illusion of impossibility. So what exactly is context?

Context is the set of circumstances surrounding a thing—a fact, a behavior, or a phenomenon in general. Let us look at an example. If we see the following two characters, 13, their meaning may be a little ambiguous; however, when they are put into context, their interpretation changes radically. Thus, in the series 12 13 14 the two characters will be interpreted as a number, and in the series A 13 C they will be interpreted as a letter. In other words, we change our interpretation because of the context. In the first case, 12 13 14, we perceive three consecutive numbers. In the second case, A 13 C, we see the first three letters of the alphabet. In this and other experiments, Daniel Kahneman and Amos Tversky demonstrated how context distorts both the meaning of the questions we ask and the answers we give.[7]

These contextual changes are also mediated by emotional responses, almost always quick and unconscious, that in the end are assessments of the extent to which our behavior adapts to different situations. The context varies in space and time and is always complex, because it includes an internal component related not only to our own condition (such as whether we are in a good or bad mood, we are sick or in good shape, and so on) but also to the characteristics of the physical world around us, the social environment in which we move, and the more abstract world of the cultural and moral dispositions that govern our behavior.

Decision-making is therefore intrinsically contextual. However, although they are not necessarily rational, contextual responses are so stable and so well adapted evolutionarily and culturally that they rarely lead to bad behavior or decisions. For instance, we do not mind lending a significant amount of money to a relative, because the relative's financial difficulties can have implications for the survival of our family. We might hesitate, however, to act similarly if the beneficiary is a complete stranger, or even simply an acquaintance. If we were to evaluate this request rationally and out of context, there would be no difference in our decision process.

In general, given these explanations, we can say that the brain is a big contextual machine: the same event, attitude, reaction, behavior, or response has different consequences depending on the context. Countless studies have shown that this process is not limited to visual perception. It applies as well to the dynamics of consciousness, which is affected by context to such an extent that the aspects of a scene to which we pay attention and that we consciously process depend on the scene's spatial context and on those things that we have attended to and processed before.

On the other hand, our brains pick up novelty immediately, so we are more likely to pay attention to what is new, in contrast to what we have already seen and experienced, which stem from known or familiar stimuli. Contrast and context are thus intimately linked, so much so that it is practically impossible to separate them. For example, in the spatial domain, anything unusual or out of place in the visual field captures our attention and directs our consciousness toward what we are attending to, leaving the rest of the scene in perceptual darkness. This kind of response is what continuously shapes our subjective experience.

And what does this have to do with magic?

A lot, because magic is an art that can take advantage of the social context. In other words, everything that happens around magicians when they are about to execute a magic trick affects how the audience members experience it and the emotion they feel at its dénouement.

In magic, the place (for instance, being on the street versus being in an enclosed space), the environment (whether it is noisy or quiet), and the type of audience (whether it is only adults, adults with children, or only children) are determining factors. In short, the same effect can be very magical or completely anodyne depending on the context. The Spanish magician Miguel Ángel Gea puts it this way:

> For this reason, the techniques work only in the context of a trick. But this goes beyond that, there are techniques that only work in smoke-filled bars, or in sessions with spectators who let themselves be transported, or for children, or for certain cultures. There are even, and this is very important to know, traps that only work in the hands of a certain type of magician, they may even only work for one magician.[8]

Thus, magic seems to be the ideal complement to neuroscience and sociology, as it helps us to advance our knowledge of how we think and act in real "ecological" situations, as opposed to laboratory conditions, which are always reductionist.

Contrast in Magic

By "visual contrast" we mean the relative difference in intensity (brightness, movement, and so on) between two points in an image. As we have seen, the brain processes only differences or changes, so if the contrast between two points is zero, the visual cells are not activated. We have also seen that the concept of contrast, as a comparison between two states or things, can be extrapolated to domains other than the visual field.

But it is also important to stress the relevance of the relationship between contrast and attention. The reality is that we process only change, and that our interpretation of the facts emerges, whether consciously or not, from comparisons. Because we need to filter the information we receive, we look only at what is useful to us at any given moment, thanks to attention. Contrast therefore plays a decisive role in attention control: a low-contrast phenomenon is something that can go unnoticed, but a high-contrast phenomenon generally draws attention.

The generation or avoidance of contrast occupies a central position in the presentation of a magic trick. During the expository phase, the magician must constantly modulate the contrast in the scene; the aim is to have the audience's attention under control at all times so that it can be led inadvertently and without suspicion to an impossible outcome. Sometimes the magician will need to capture the audience's attention during the expository phase of the trick by creating a significant contrast between two elements, whether physical or part of the story line. On many more occasions, the magician will need to achieve the opposite effect—by not attracting attention, the magician will avoid contrast and the secret maneuvers of the trick's internal life will go unnoticed.

Magicians, when they present their effects, are true masters, continuously combining actions to attract attention and avoid contrast. In the following sections, we use the word "contrast" to refer to both situations of attention control: when attention must be attracted, increasing the contrast, and when attention must not be drawn, thus avoiding contrast.

Avoiding or Reducing Contrast in Magic

The inability to process all incoming information in real time has made our brains behave like extraordinary machines capable of predicting events with a high degree of reliability based on our experiences. For this reason, when what is observed concludes as predicted, adult spectators relax their attention and discard anything supposedly superfluous. In magic, the artists try to carry out their maneuvers in a logic of "normality" that makes everything predictable, because "what matters is what the spectator feels, not what he sees."[9] On the contrary, if the predictions are not fulfilled, alarms go off immediately; the brain then needs more time to process the data, and there is a risk that its attention will be focused on the failure of the predictions. Such a disruption is fatal, because "if the audience feels compelled to analyze why you did something, you've already lost the battle."[10]

In short, the central objective of the magician's presentation is to lead spectators toward the climax without giving them a chance to stop to question anything, a goal that is achievable only if throughout the external life of the trick there is no undesirable contrast.

This objective requires, of course, that each and every part of the expository phase be aligned for this purpose: the construction and structure of the magic trick; the techniques, materials, and methods to be used; and everything related to the setting of the presentation. Illusionism is the *ars artem celandi*, the art of hiding art. In the words of the Spanish magician Gabi Pareras, "the trick must be like a crystal figure inside a glass of water, which cannot be seen, felt or sensed."[11]

General principles of good practice have been consolidated in the world of magic, and many of these principles have a common denominator: avoiding undesirable contrasts. These are principles of logic and predictability that, when applied during the presentation of the effect, leave spectators in the best position to be shocked by the trick's unexpected outcome. The logic and predictability of the presentation are indispensable for manipulating the brain's ability to infer and anticipate the facts, avoiding any undesirable contrast.*

Strategies and Resources during the Presentation of a Magic Trick

In magic, nothing is subject to improvisation, even if it seems so. Magicians create opportunities that may seem fortuitous to us, but these apparently improvised opportunities are the result of many previous trials. At the same time, thanks to their resources and experience in this kind of situation, magicians also know how to take advantage of opportunities that may arise casually. What are the qualities and tactics on which a trick's presentation relies?

First, clarity is fundamental: the trick must be presented in a clear, simple, and understandable way. Everything the magician explains and does must be received as natural. Let's take an example: In rope magic, magicians strive to tie the knots as anyone normally would. Not only must the method and the handling of the materials be totally unnoticed, but the rhythm of the exposition and the "timing" and execution of the maneuvers must also be appropriate. Any mismatch in the timing of the movements, for example, would arouse suspicion. Further, the presentation must be constructed in such a way that the trick is easy to understand, with a thread that can be easily followed. It is important to bear in mind that verbal instructions can be inherently deceptive.[12] Moreover, a presentation is confusing unless it is consistent and avoids moments that are not easily understood—that is, that are not followed smoothly—in one go.

*As Gea comments in *Numismagia & percepción*: "The brain has many mechanisms, but we mainly take advantage of the fact that it constantly assumes and predicts. We have to find the balance so that it neither assumes nor predicts more or less than is good for us, that it is aware of what is right and does not get carried away more than it should. Our goal is to get it to assume what we want, to predict what is convenient for us, and to have its analysis overcome by the beauty of the experience."

Fatigue is another factor that creates distraction, causing spectators to disconnect. When presenting a trick, good magicians make it easy for spectators to follow along so that they do not become exhausted; as all magicians know, you should not ask the audience to remember *several* cards. The American magician Simon Aronson notes that a "spectator must first be convinced that he is aware of all that has happened: that he has been attentive, that he's followed everything, that nothing has escaped his notice."[13]

By the same token, naturalness is key: the gestures, maneuvers, and actions of the magician during the trick must be completely natural and fully justified, even when he pretends to be manipulating nonexistent objects, because any strange, unusual, or unjustified movement will raise alarms.[14] Spectators will take notice and be distracted from the logic and predictability of the presentation. "Preventing the audience from questioning something is far more effective than having to convince them later that something has not happened."[15]

Let's take another example. A magician who goes onstage and starts by saying, "I have a normal rope," will be going against all the principles mentioned earlier. If the rope is normal, why mention it? And if the rope is *not normal*, then it is better to be misleading and keep silent.

As for naturalness, this must be a quality not only of magicians' gestures, maneuvers, and actions but also of their attitude. Imagine, for example, a magician using a tidily "stacked" deck, with the cards arranged in an order that only the magician can understand and remember. If the magician is relaxed and carefree, the spectator will not question the deck, especially if it is left on the table near the spectator. When he places the "stacked" deck squared up in front of the spectator, the Spanish magician Dani DaOrtiz tilts his body back without moving from the seat; he angles away from the deck but makes no move away from it, which would be unnatural.[16]

Sometimes magicians take advantage of a justified maneuver, or of a certain maneuver that has its own logic, to make a secret move. We are referring to what Ascanio called "actions in transit," such as taking a pen from your pocket to write with and taking advantage of this maneuver to carry a coin. As the American magician Al Baker puts it,

"Actions that appear necessary but unimportant are only half-noticed and soon forgotten. Actions that are unnecessary arouse suspicion."[17] Necessary, seemingly unimportant maneuvers profit from the fact that we make unconscious inferences about the logic of a movement or an act, and these inferences nullify any conscious deliberation about the action's motives.[18]

Arturo de Ascanio developed, using his own grammar of cognitive metaphors, a catalog of techniques and recommendations to avoid undesirable contrast throughout the expository phase of a magic trick. As he used to say, the "perfect technique is the one that isn't there." He was very insistent that the story line must be clear and self-evident, that all gestures and acts made by the magician must be justified and have a purpose. "What is deceiving is the ease."[19] The best proof that the magician has succeeded is to witness the typical reaction of the audience at the end of the trick: "No way. How did he do it?"

There are some complementary techniques in magic that contribute to the logic and predictability of the magic trick. These techniques are intended to enable the magician to avoid undesirable contrasts at certain moments or maneuvers during the presentation. Among them, we can identify three: "conditioned naturalness," ruses or feints, and illusory correlations.

Conditioned naturalness, a concept coined by Ascanio, refers to a kind of conditioning during a brief length of time when the purpose is to normalize, always by repetition, what in any other context would attract attention. This technique is used when the magician needs to perform maneuvers that are rare or unusual. Because our brains work in a statistical way, these maneuvers would present great contrast and powerfully draw our attention. To avoid this, magicians resort to repetitions that normalize a movement or maneuver, thus reducing the contrast created by an unnatural manipulation.

Consider a magician attempting to accustom the audience to an infrequent or atypical way of picking up a deck or dealing the cards; although unusual in any other context, this maneuver ends up looking natural by mere repetition. The Italian-Argentinean magician Tony Slydini brought some personal subtleties to his tricks based on this technique. In one of them, Slydini constantly raised and lowered his hands near the edge of the table. Once the spectators got used to this movement, the magician could drop anything into his lap without an interruption in what had become his normal movement pattern.[20]

Sometimes magicians manage to make unusual maneuvers in the background during the presentation of a trick. The overall choreography is what creates a sense of naturalness, making it easy for these unusual maneuvers to go unnoticed.

Ruses, feints, deceptions, simulations—all serve to make you believe one thing when in reality something else is happening. The first phase of a ruse is always preconditioning. In coin tricks, for example, the first time a magician passes a coin from one hand to the other, she does so without cheating; that is the preconditioning action. But then, when she repeats the action, she adds a trick, such as a false transfer of the coin to the other hand the second time, resulting in its alleged "disappearance."

Finally, illusory correlations are associations whose purpose is to reinforce the logic of a trick. Magic takes advantage of the natural tendency of humans to infer cause-and-effect relationships in situations where there is only a correlation. For example, if B (rain) always appears after A (clouds), we assume that A is the cause of B.* Because illusory correlations appeal to our tendency to interpret the world in terms of cause and effect, they have the great advantage of not producing contrast. They can add consistency and support to a given story or narrative.

In coin tricks, magicians very frequently use illusory correlations to convince the audience of the truth of multiple passes. For example, to reinforce the illusion that there is more than one coin in play, whether in one hand or some other place, the magician manages to produce the noise of the supposedly multiple coins coming into contact with each

* The existence of a correlation between different phenomena does not necessarily imply causality. There are many surrealistic and spurious correlations, such as "US spending on science, space, and technology correlates with suicides by hanging, strangulation, and suffocation" (see Vigen, "Spurious Correlations"). Such spurious correlations lend credence to the beliefs in coincidence held by those who are convinced of the existence of so many conspiracies because "everything is interconnected." Conspiracy theories proliferate when statistics are used literally.

other. (This is how the classic “Coin Click Pass” works.) In the routine known as the “Miser’s Dream,” the magician produces coins out of thin air or from the ears and noses of spectators.[21]

In short, most strategies used by magicians are ultimately aimed at avoiding the production of undesirable contrasts. This general objective—avoiding contrast in the expository phase—is what Ascanio so poetically described as the “magical atmosphere attained only when all factors come together simultaneously in perfect harmony.”[22]

Presensory Manipulations

Many materials and methods used in magic conceal and interfere with stimuli in a presensory stage—in other words, before these stimuli can activate any nerve cells. In such cases, the information has already been manipulated when it arrives at our sensory organs, a manipulation that the entire audience experiences equally. For example, whatever is physically hidden cannot be seen by anyone, period.

Magic makes use of concealment almost continuously, and it has learned to do so in many ways, not only with the help of various gadgets and gimmicks but also sometimes by resorting to new materials or to other sciences, like optics, and even by using auditory masks, such as a certain timely bit of music or an appropriate noise to cover up the sound made by a secret gadget. The simplest examples of concealment are those based on camouflage, which reduce to zero the contrast of objects with the background and render them invisible—as in black light theater, or when magicians use cards with a black back on a black mat.

Another effective way to manipulate what the audiences perceive at a presensory level is to use optical illusions. As a cautionary note, in this book we distinguish three kinds of illusions: optical, visual, and cognitive. Optical illusions are those interferences with perception derived from the physical properties of light. They can originate, for example, from the use of structural devices that distort perspective or from using tools (smoke and mirrors) that produce reflection or refraction of light. A classic example of refraction is the pencil that appears to be broken inside a glass of water. The important thing about optical illusions—as opposed to visual or cognitive illusions—is that they are produced from a previous physical alteration. In other words, what reaches our sensory organs is already distorted when we receive it. They are presensory illusions. Optical illusions occupy a central place in the magic of "grand illusions" (stage magic with large objects), whereby black spaces, mirrors, and other gadgets are used to hide objects and distort size when seen from a certain distance.

The magic of "grand illusions" benefits not only from optical illusions but also from the use of visual and cognitive illusions, as when depth perception is altered using illusory perspectives. Unlike optical illusions, visual and cognitive illusions are not presensory and can therefore be perceived in subtly different ways by each spectator. These illusions derive from various strategies that the brain uses to process information. In the later chapters that focus on attention, memory, and decision-making, we present some examples of visual and cognitive illusions, such as bent spoons—which combine elements of visual illusion with other elements more typical of cognitive illusion—and "forcing," a cognitive illusion in which spectators falsely believe that they have freely chosen a card from all the cards in the deck.

CHAPTER 6

We Filter and Process Only What Is Useful to Us

The Attention Filter

To efficiently manage the flow of information, whether captured externally by the sense organs or generated internally by our own thoughts, the brain has at its disposal a great cognitive process: attention.

Attention is the process by which certain information is filtered or selected from the many sensory stimuli we constantly receive.[1] It helps us filter and discard irrelevant information, allowing us to focus on what stands out the most and select only what is useful at any given moment. Only then can selected information be processed—that is, perceived.

Through attention, we focus our senses and analytical capacity on a certain point, object, or act—including on our thoughts. At times, we may be paying attention to something other than what we are looking at.

We can distinguish between two main types of attention: overt attention, when we look directly at the object of our attention, and covert attention, when we look at something other than the object of our attention. In other words, we can direct our eyes to an object to attend to it, or we can fix our attention on it without moving our eyes to it.

Attention and Awareness

Attention, short-term memory, and awareness are closely related systems in that neither attention nor awareness would be possible without short-term memory. Information that we attend to remains for a while in short-term memory, where it is used to elaborate thoughts or to make decisions.

Although in this book we focus on the attentional phenomena of which we are aware, it is important to note that most attentional mechanisms operate unconsciously; that is, many forms of attention occur automatically without our being fully aware of them and we know only the results. This fact is also important in magic, as we will see later.

Our attention system evolved to be selective. Attention expresses our "logistical" ability to allocate limited brain resources to different tasks over time and space. Not all information in our world is equally relevant to us, nor has it ever been for our survival. For example, from the first years of life, we are conditioned to react to our name or surname to such an extent that, pronounced in unusual contexts, on the street or in a restaurant, it will immediately catch our attention. In this regard, Colin Cherry has described our ability—familiar to most of us—to listen to only one of the many conversations that take place at a crowded party, even if it is not the one we are involved in, as the "cocktail party" effect.[2] This ability to filter our attention allows us to select one of many conversations and bring it to our awareness amid all the background noise. Moreover, even if we are attending to one conversation, a good chunk of the rest of the information in the environment is still temporarily—and amazingly—retained in short-term memory so that we can recover it if our attention is interrupted. This is why we can detect our name being spoken amid multiple conversations: our brains have learned to separate the grain from the chaff, to process only the information that, in the past, was useful to us in understanding the world, and to practically ignore everything else.

Our attention is "awakened," or made aware of a contrast, when new information appears—when what we observe has no equivalent in the memory, when it differs from the logic of the facts we have been constructing, or when an exceptional situation occurs. A known or familiar piece of information, such as the arrangement of the living room furniture in a friend's house, will often not attract our attention. But it will be awakened on our next visit if a piece of furniture is in another place, or if we voluntarily and attentively look for an object in the environment.

Attention can also be actively captured by sudden and unexpected stimuli (appearances, noises, lights). Indeed, this is one of the main strategies used by magicians to attract and control the attention of their audience. There are some exceptions, but they confirm the rule. For example, when some of these unexpected stimuli occur in daily life, we do not necessarily always attend to them. The same happens if they occur while we are deeply concentrated on a specific task. The attention of a car driver arriving at an intersection in a city with very dense traffic is attentive to what other cars are doing while deciding when to cross. Unfortunately, with her attention so focused on cars, she may not see the cyclist crossing in front of her. Our capacity for selective attention—for selecting the unexpected stimulus to which we should really pay new attention—operates constantly on the margin of conscious perception. In other words, selective attention is paid by the unconscious brain, which, as we will see in the following pages, assists us on an ongoing basis.

The Concept of "Misdirection" in Magic

Magicians coined the term "misdirection" to refer to one of their best tools, specifically, attention control. Misdirection has appeared in books for magicians since the early 1900s, including those by the legendary Nevil Maskelyne and the American Harlan Tarbell, who, according to the American magician John Mulholland, used the word "misdirection" long before anyone had heard of the term "psychology."[3]

But what does "misdirection" really mean?

The word is defined as "diverting" or manipulating attention through actions that make the viewer avert his or her eyes from a specific place—namely, from the place where the magician needs to make maneuvers to execute the magic trick.[4] The misdirection always begins by capturing attention and then redirecting it to a new place, dividing it between two or more targets, or either relaxing or completely deactivating it.[5] Over time, however, the term "misdirection" has come to refer to any manipulation aimed at controlling attention, such as covert deviation when the magician manipulates your attention without necessarily manipulating the direction of your gaze.[6]

In turn, other magicians have revisited the concept, emphasizing that "misdirection" refers more to the art of directing attention than diverting it. As suggested by the Dutch magician Tommy Wonder in his excellent critical analysis of the subject, misdirection is a technique intended not so much to divert attention from something secret as to direct it toward something interesting: "It would be far better for us if *misdirection* had not become an accepted term in magic, and *direction* had been adopted instead."[7] According to Wonder, if magicians are not continuously directing the audience's attention, they risk losing the audience's attention to something unrelated to the act.[8] More recently, different definitions and classifications of misdirection have been proposed from a purely psychological perspective that deviate somewhat from this classical view.[9]

Focal Attention

We only stop paying attention naturally during sleep or in situations of induced or pathological coma. Therefore, in the waking state, it is highly unusual for the conscious brain not to be constantly focused on something, whether attending to internal issues resulting from our stream of thoughts or attending to the external stimuli forever present in the world around us. In fact, when we are awake, it is so difficult not to be focused on something that to "switch off," we need special training similar to learning how to meditate.

Attention is not uniform across all spaces. That is why, when attention is oriented toward a particular place, it is called focal attention. The smaller the space one attends to in a focal way, the more efficient the processing of the information obtained and the fewer the cognitive errors.

When attention is persistently focused on a very specific point, a decline in attention outside that focal point inevitably occurs. In other words, if we focus on a certain area, everything around it will be progressively neglected to such an extent that we will ignore movements or maneuvers made very nearby. When you are very concentrated on studying a document or reading an engaging book, it is very easy for someone to enter the room and even walk by your side or talk to you without you being aware of their presence.

This state is what is known as "suppression of the periphery," a phenomenon that magicians have used since time immemorial to perform necessary maneuvers in areas to which spectators are not paying attention.

Focusing our attention, however, does not guarantee that we will perceive everything that is happening at the focal point, as Tim Smith, Peter Lamont, and John Henderson have shown in coin and card tricks, as have Gustav Kuhn and his colleagues.[10] Focusing always requires a competition between our train of thought (endogenous capture) and external events or places that attract our attention (exogenous capture), which leads as a secondary consequence to divided attention. Let's take a closer look at what that involves.

Exogenous Capture of Attention and Open Diversion

The exogenous capture of attention and the subsequent open diversion, also known among some magicians as "physical misdirection," is one of the classic methodologies of magic. Magicians create new areas of attention with their body language, gestures, and gaze so that they can perform their secret maneuvers outside the area of interest.

Inside our brains, an open diversion of attention interferes with the perception process because it prevents sensory information from entering short-term memory. That makes further processing more difficult.

In magic, there are several methods for exogenous capture and subsequent open diversion, among which three in particular stand out. Let's examine them one by one.

Contrasting Stimuli

Contrasting stimuli present one of the most effective methods of attracting attention. Based on the creation of sudden and unexpected change, they generally work by causing surprise. Loud sounds, bright lights, changing contours, and unpredictable movements in the peripheral field of vision are all stimuli to which we react immediately—and their effectiveness will be directly related to the degree of contrast produced. In other words, the relevance of a stimulus depends on the context in which it occurs.

For instance, a naked person on a nudist beach would not make an impression, as they produce no contrast. However, if that same person appeared in the middle of a meeting of the US Congress, they would certainly grab attention. As we saw when discussing the visual structure of a scene, the points of maximum contrast, the areas of maximum rate of change, both spatial and temporal, are those that capture the most attention. In magic, the most commonly used contrasting stimuli are sounds (such as changes in rhythm, tempo, or other musical punctuations) and surprise appearances: the rabbit coming out of a hat or the sudden change of color in a handkerchief.

It is also true, however, that, if the stimulus that has captured attention occurs repeatedly, it will become predictable and lose contrast, and even though the stimulus remains just as strong, its ability to attract attention will decrease. In the 1800s, this principle was used to train warhorses to withstand loud and unexpected noises around them, such as firecrackers, so that they would not abruptly react in battle.

Nonverbal Communication in Magic

You never get a second chance to make a first impression.

In magic, nonverbal communication is a fundamental part of presentation, as it is strategically essential for attention control. Nonverbal communication refers to the set of signals given off by our bodies through which we interact and relate, such as our facial expression of emotions, our body language, and how we navigate personal space in our interactions with others.

Nonverbal communication is part innate, part imitative, and part learned. During the 1950s, the psychologist Albert Mehrabian popularized the rule that more than 90 percent of interpersonal communication that includes the communication of feelings is nonverbal. Fifty-five percent of this nonverbal communication is body language; the tone, pauses, and intonation of the voice make up 38 percent; and the content of our speech amounts to a mere 7 percent.[11] Although these percentages are, of course, imprecise measurements, they highlight the importance of all nonverbal aspects of interpersonal communication.

The face and its elements (eyes, eyebrows, mouth, wrinkles) provide the most reliable expressions of a person's mood and emotions. When we want to know how someone is, the first thing we do is look at their face; as the saying goes, "the eyes are the mirror of the soul." In facial expression, the gaze carries the greatest weight of nonverbal communication, and for that reason, it is a key strategic tool in magic. Magicians regulate the direction, intensity, and duration of their gaze not only to reveal where their own attention is focused, but also to better understand their audience and to deliver implicit instructions to attend to one object or another.

In addition to gaze, body language is another crucial element of nonverbal communication, in particular posture and gestures. Posture eloquently conveys feelings, attitudes, and aspects of one's personality, particularly as posture cannot always be tightly controlled at a conscious level. A gesture is posture in motion. It is said that a language is not only spoken but also gestured. The movements of the body—the hands in motion being particularly relevant in magic—contribute to the communicative act since gestures almost always accompany and reinforce words or actions.

Respecting personal space is also relevant. When spectators are invited to participate in magic tricks, the distance they keep from the magician is set unconsciously and can vary greatly depending on the culture. Pickpocketing experts in magic have learned to manipulate distances

FIGURE 6.1.1. "Misdirection Trick with Cigarette."

from their "victims" in a show, approaching them from the side so that their personal space does not feel directly invaded.

High-Interest Zones

Nonverbal communication is such a powerful mechanism for attracting attention that it can even prevail over the magician's explicit instructions during a trick.[12] The Swiss psychologist and magician Gustav Kuhn has studied the effectiveness of exogenous capture during a trick in which the magician makes a cigarette and a lighter disappear in view of the spectators. In these tricks, performed live, magicians capture the audience's attention particularly at the moment when they pick up the cigarette and put the wrong end into their mouth while lighting the lighter. (See the video "Misdirection Trick with Cigarette" at figure 6.1.1.) This is how so-called high-interest zones are created. Such moments capture and divert attention so that magicians can perform unattended maneuvers—in this case, dropping the cigarette and lighter into their lap, a motion unseen by most spectators.

Magicians draw the audience's attention toward the zone of interest with consecutive uses of surprise (the sudden disappearance of the lighter), their body language (the direction of their gaze and their body movements), and other contrasting stimuli, such as, in this case, hand gestures accompanied by the sound of snapping fingers (figure 6.1.1). What these Kuhn studies confirm is that, in the end, it is the audience's attention that is manipulated, not where they look.[13] In fact, subsequent studies have shown that the direction of the gaze is not as relevant;

many spectators discover the trick using their peripheral vision, even if they mistakenly believe that they have looked directly at the high-interest zone.[14]

The Power of Nonverbal Communication

In his laboratory, Gustav Kuhn also studied the effect of the vanishing ball. Kuhn wanted to prove that the success of the trick depended on the magician's nonverbal communication (see videos "Vanishing Ball Illusion" at figures 6.1.2 and 6.1.3). These were the results: When the magician performed the necessary social signals and directed the audience's gaze, with the support of arm and hand gestures, 68 percent of the participants were able to "see" the fake throw in the last attempt. However, when the magician did not use these signals, only 32 percent were able to see it.[15]

It is probably the consecutive repetition of the same maneuver that explains why viewers anticipated seeing the ball again—even anticipated seeing things that never actually happened. This anticipation has been called "verbal suggestion" by some authors.[16]

Other studies have observed that, although such social signals can condition an audience's expectations with respect to an action or object on stage, not all audiences have the same sensitivity.[17] For example, if some spectators have a problem focusing their attention, such as people with autism, the effectiveness of social cues in achieving magical effects may be altered.[18] Given the heterogeneity of this spectrum of conditions, further research is needed.

Effectiveness also depends on the type of trick used.[19] Contrary to what was observed by Kuhn's group in the vanishing ball trick, where nonverbal communication seemed to be the determining factor, Susana Martinez-Conde and her colleagues, in their study of coin tricks based on fake and real tosses, found that their effectiveness was greater when the magician's head was covered—that is, when spectators were not looking at the magician's face at the crucial moment. The authors concluded that, in some cases, social clues could be redundant or even detract from the power of magic.[20]

FIGURE 6.1.2. "Vanishing Ball Illusion: Pro-Social Cues Condition."

FIGURE 6.1.3. "Vanishing Ball Illusion: Anti-Social Cues Condition."

These contradictory results could indicate that each type of trick uses social cues or nonverbal communication differently. However, these contradictions can only be resolved through further research.

Managing the Gaze

We have already emphasized that the gaze of the magician is an essential tool throughout the presentation of a trick. Good management of the gaze can determine the success or the failure of a magic act. Spectators look at magicians' faces and follow their eyes: if the magician looks at a given point—sometimes reinforcing his gaze with the appropriate body language—spectators will also look at it. Take a trick starring Miguel Ángel Gea, in which we tracked where spectators were looking. We analyzed two versions of the act. In one version, Gea looked at the audience while he concealed the coin in his hand (figure 6.2, upper row). In the other, he looked at his hands while he was concealing the coin. When the spectators followed Gea's gaze to his hand, they detected the trick more often (figure 6.2, bottom row). The audience looks where the magician looks. Controlling gaze and using other nonverbal communica-

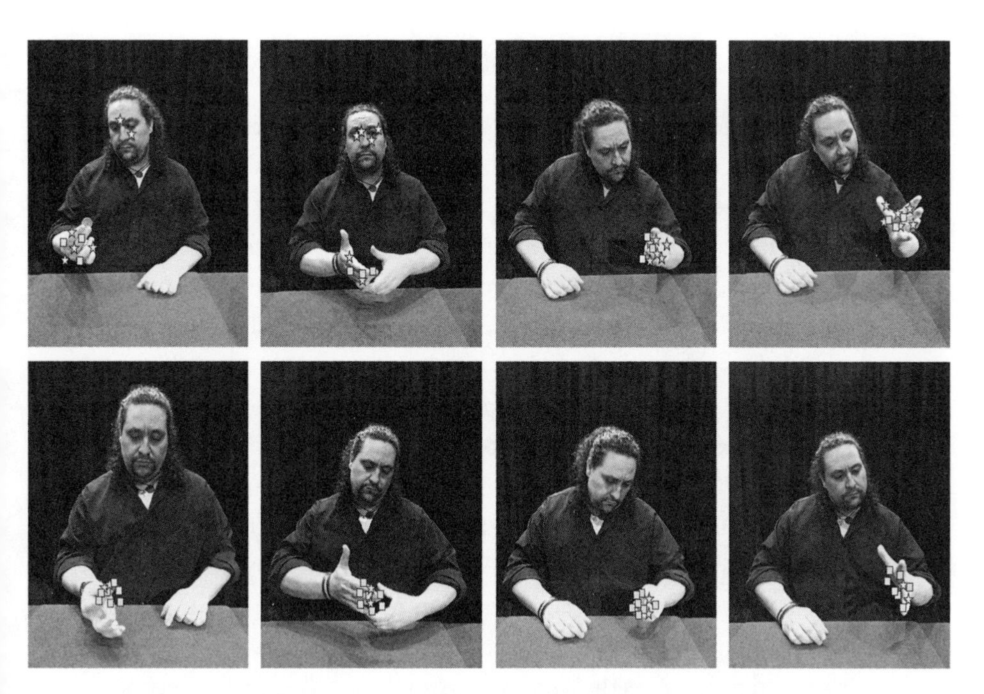

FIGURE 6.2. Miguel Ángel Gea conceals a coin by means of a false transfer of the coin from one hand to the other. The marks (stars and squares) indicate where the spectators are looking at each moment. Stars mark the gaze of those who did not discover the method. In the top row, second image from the left, when the magician looks at the audience, he deceives only those who follow his gaze.

tion strategies can clearly be very effective, especially when combined with other open diversion strategies.[21]

Finally, along with gaze and other social signals, magicians capture attention through implicit instructions—that is, through instructions not expressed verbally and transmitted by nonverbal means. These instructions can manage expectations with stereotypical actions or even lures, such as when the magician reaches out with his hand toward the spectator instead of directly asking her to give back the deck that she is holding. The Canadian-American magician James Randi noted that it is easier for viewers to accept implicit suggestions than direct assertions because the latter arouse suspicion: "Is the magician trying to fool me?"[22]

Priority Movements

Not all movements have the same capacity to attract attention, and magicians make good use of so-called priority movements to capture the attention of their audience.

First, a large or fast movement can cover a small one. In neuroscience, an equivalent of this effect would be the "motion silencing awareness" illusion, in which static points that constantly change color seem not to change colors when they also move rapidly (see "Motion Silences Awareness of Color Changes" in figure 6.3).[23] Two theories have been proposed to explain our inability to detect these types of changes. The first, known as the "temporal freezing" hypothesis, postulates that our perception is anchored in the initial condition and we are not aware of the changes. The second, called the "implicit updating" hypothesis, proposes that we detect each new situation of the stimulus but are not aware that nothing has changed with respect to the previous situation. This second proposal is the one that currently has the most experimental support.

Second, when several movements are performed simultaneously, spectators tend to follow the one that starts first. The magician Arturo de Ascanio called it the "law of priority movement."[24] When there are two simultaneous paths of movement, the effect is greater if the beginning and ending movements are slow.[25]

Third, it has been suggested by the American magician Apollo Robbins that a curved movement attracts more attention than a straight one.[26] The reason is that when a movement is done in a straightforward manner, the observer's attention is mainly, if not only, focused on the initial and final parts of the movement. Susana Martinez-Conde and her colleagues have found that, in fact, during such movements the eyes of the spectator make a large amplitude movement, a saccade straight from the beginning to the end that temporarily disconnects attention.[27] When the trajectory is curved, however, the eye makes no such large saccade; the gaze—and hence the attention—of the audience continually follows the magician's hand as it moves, leading to a more surprising magic effect. Other authors have reached contradic-

FIGURE 6.3. "Motion Silences Awareness of Color Changes."

tory conclusions on curved movements.[28] Their contribution to the discussion has been to show that it is not just the trajectory that matters for the control of attention but the location and direction in which each movement is made.[29]

Endogenous Capture and Covert Deviation

In contrast to exogenous capture—being attracted by external stimuli—we can also direct out attention at will. This is called endogenous (active or "top-down") attention: we can direct our attention when we voluntarily select where and when to focus—for example, when we are waiting for the announcement of our boarding gate on the loudspeakers of an airport, when we identify a particular person in a crowd, or when we look for a taxi amid the traffic. In these cases, we prepare by focusing our senses on the place and time of the event we are expecting.

Exogenous attention gathering (attracted by external stimuli) is quicker than endogenous attention capture, although more ephemeral. Exogenously captured attention does not remain oriented to the place of attraction for long, unless, as in magic, the magician is trying to capture the audience's attention permanently by continuously refreshing the focus of attention. Attention captured endogenously, on the other hand, can be sustained for longer because it is oriented toward an expectation, an objective of interest, or a particular train of thought. Endogenous attention can be interfered with by a stimulus that changes the focus, although it is equally capable of being sustained in the presence of any interference if the value of the signal expected is sufficiently important. For instance, when you are hungry and waiting to be assigned a table in a restaurant, any stimulus around you goes completely unnoticed as all your attentional resources are directed to the expectation of hearing your name.

Endogenous capture is linked to covert diversion (or psychic "misdirection" in magic). Covert diversion is achieved when the focus of the spectator's attention is shifted to another subject, fixing the spectator's thought on something concrete without necessarily having to mediate a change in the gaze ("you're looking, but you don't see"). As we will see next, in magic, dividing attention is the most common psychic "misdirection" maneuver.

Divided Attention

> I get ready and, out of the blue, I say to a spectator sitting in front of me, "Can you count to sixty?" The spectator looks at me, dumbfounded, not knowing how to take the question. Others look at him smiling. The whole thing lasts only a second, which is more than enough time for me to look at the card.
>
> —Alfred Binet, "La psychologie de la prestidigitation," *Revue Philosophique de la France et de l'Étranger*

In a 2010 study about divided attention outside the laboratory, researchers had a person dressed as a clown ride around a city park on a unicycle. About 60 percent of the passersby who were strolling with a companion noticed the clown. However, only one in three pedestrians who were listening to music or just walking around mentioned seeing him. What is interesting is that, among those who were talking on their cell phones, only 8 percent spontaneously recalled the presence of the clown. The response rate increased in all cases when specifically asked whether they had seen the clown on the unicycle, but still, only 25 percent of the phone users recalled seeing the clown.[30]

As this example illustrates, attending to two tasks at the same time can be very difficult, as difficult as having two conversations at once. For instance, if we are asked to detect certain changes in two audio or video files simultaneously, we are very likely to not succeed. We fail because our attention is limited by our short-term memory capacity. We can easily attend to one thing after another but are unable to attend to several things at once. In the latter case, short-term memory becomes saturated and our ability to pay attention becomes inefficient. Like trying to empty a bathtub full of water with only a glass, we cannot do it all at once.

To minimize our inability to consciously process two inputs at once, the brain builds an illusion of simultaneity. It can do this because attention shifts between one task and another with such extraordinary speed (every 250 micro-seconds) that we are not aware of the changes.[31] Therefore, contrary to popular belief, there are no true multitaskers, but only people so well trained in certain tasks that they can perform them sequentially with extraordinary efficiency.

Let's look at some cases. It is not very hard for you to perform a well-learned activity—one stored in the implicit memory, such as riding a bicycle—while answering specific questions at the same time. You can peel an apple or talk on the phone with a relative while watching a favorite television show. But you cannot do a crossword puzzle or interpret a feature article while texting a friend because each activity demands much more attention. By the same token, if you want to enjoy a magic act, you cannot be engaged in other activities at the same time.

When multiple demands are being made on your attention, you make more mistakes. In addition, when the mind is busy, the delay caused by other stimuli, usually a third of a second, increases, contributing to our inability to attend to two stimuli at once. For instance, we can listen to music while driving a car, but we become distracted when checking the GPS or attending to a mobile phone. Numerous studies show a higher rate of accidents and violations under the latter circumstances, and most countries around the world have now banned talking on the phone (or seeing and responding to text messages) while driving.

Because cell phones have become a kind of permanent human prosthesis, this risk has extended to driving other types of vehicles and even to just walking on the street. In the Netherlands, not surprisingly, texting while riding a bicycle was banned in 2019, and in Honolulu it is now illegal to text or even look at a phone while crossing the street. When you divide your attention between different sets of information you receive, you miss out on things. This phenomenon has been labeled "inattentional blindness," and as already mentioned, it can have adverse consequences.

In 1999, the scientists Daniel Simons and Christopher Chabris published a highly acclaimed video of an experiment involving students playing basketball. One team was dressed in white and the other in black. The spectators were asked to count the number of times the players on each team passed the ball to each other. During the game, a person wearing a gorilla costume jumped onto the field, walked among the players, and finally disappeared. At the end of the game, 58 percent of those counting the passes of the team dressed in black, the same color as the gorilla, were able to detect the gorilla, while only 27 percent of those counting the passes of the team dressed in white had noticed it. This experiment showed that the task of counting the number of times the students passed the ball was sufficiently absorbing that many did not notice an obvious detail of the scene (see "Selective Attention Test" in figure 6.4.1). The difference in the results for those counting the passes of the black team and those counting the white team's passes stems from the fact that, for the former, the gorilla matched its target color, while for the latter it did not and simply formed part of the background. A few years later, in 2002, the same authors published another version of the video with additional details that went unnoticed even by people already familiar with the original example (see "The Monkey Business Illusion" at figure 6.4.2).[32] In short, the more we divide our attention, the less information we can extract.

So, in magic, if we are reading text messages while being introduced to a trick, we are likely to miss the effect. A perfect example is the video starring the British magician Derren Brown, based on an original experiment, also by Daniel Simons, that combines change blindness (which we saw in chapter 4) with inattentional blindness. Brown appears on a street as a tourist, map in hand, asking passersby to explain how to get to a city landmark. In the middle of the explanation, a visual interruption occurs, with surprising effects (see "People Swap—Derren Brown" at figure 6.4.3).

Dividing attention is one of the most effective and frequently used techniques in magic tricks. It is used both to hide methods and to make

FIGURE 6.4.1. "Selective Attention Test."

FIGURE 6.4.2. "The Monkey Business Illusion."

FIGURE 6.4.3. "People Swap—Derren Brown."

it difficult to reconstruct the trick. As we will discuss in chapter 8, divided attention not only compromises the capacity of short-term memory but also affects memory formation (the passage from short-term to long-term memory) and interferes with evocation and the ability to remember.[33]

In our neuroscience classes with magicians, to illustrate the concept of divided attention, we perform an effect that we call "The Busy Brain," based on ideas of the legendary Austro-Hungarian magician Hofzinser. It consists of distributing several decks of cards among the audience so that they look for a card that they have chosen previously. The search must be done by turning the cards of the deck faceup, one by one, while counting to determine in which position the card appears. The trick ends when the whole deck is turned over without the chosen card appearing, as if the magician knew which one it was and made it disappear. The reality is that the spectator is forced to choose an even card, while the decks given to the public only contain odd cards. The demand to recognize the card and determine in which position it appears is so strong that spectators never realize that there are no even cards in their deck nor that the chosen one was even.

The most common techniques used by magicians to divide attention are sudden distractions and the provocation of very demanding tasks, such as in the example just given. Sudden distractions can be achieved by asking questions. In magic, Ascanio's "obfuscating question" stands out: "Do you believe in small miracles?" These are questions that have no relation to the effect and that serve to interrupt the thread of the story and divide the attention of the audience for a few seconds, enough time for the magician to perform a maneuver in an unnoticed way that, under normal conditions, would have been very evident.[34] Juan Tamariz updated Ascanio's obfuscating question by increasing the number and complexity of the questions to save time for performing secret and more complex maneuvers.[35]

An extreme example of divided attention achieved by assigning highly demanding tasks is that of every "victim" of a pickpocket artist. First, the pickpocket artist overwhelms the victim with speech and movements. Then, while the pickpocket artist takes something away, they stare at the victim directly in the eyes. This provokes diverse and simultaneous focuses of attention, and the victim is unable to attend to both. In addition, the pickpocket artist often invades the victim's personal space (not aggressively, of course) and distracts them with incessant patter. This stratagem not only divides the attention of the victim but also practically cancels out their capacity for attention, because the victim is not able to follow the multiple and simultaneous actions of the artist. In these circumstances, the pickpocket artist can remove the victim's watch, tie, glasses, or belt without the victim noticing, because, in addition to shifting and dividing the victim's attention, these maneuvers provoke tactile post-sensations (as we have already seen, for example, when removing a watch unnoticed). The technique relies on the same principles as those of real pickpockets. The street pickpocket, besides being very skillful with their hands, divides the victim's attention by giving him or her a seemingly innocent push or shove and taking advantage of that moment to steal a wallet or cell phone.

Temporary Control or Continuous Direction of Attention

Good magic without proper attention management is an impossibility.

—Tommy Wonder

Attention is not homogeneous in space and does not remain constant over time. Control, anticipation, and modulation of temporary attention depend on basic mechanisms (eye movements, brain rhythms, and so on) that are not yet very well understood.[36] We know that the audience's attention fluctuates throughout a show, seeks moments of refreshment, and may be subject to different distractions. In these circumstances, the magician's objective is to temporarily appropriate the audience's attention, to control it during the entire expository phase of the magic act, and to ensure that, during this period, spectators do not stop to "think" but rather remain attentive to the presentation and do so automatically. Taking temporary control of attention in a magic act is an objective in itself, not only to prevent the audience from noticing what is not appropriate but also to ensure that they follow the whole expository phase so that no contrast is produced. If the magician's temporal control of attention fails, the audience becomes distracted, fails to follow the magician, and the trick fails. The advice from the magician Darwin Ortiz is particularly apt: "Never let the audience slip through your fingers."[37]

To maintain temporary control of attention during the expository phase, the magician's narrative must achieve a solid thread that is followed by the audience. As we have underscored, the trick's story line, or narrative, and the simplicity and clarity of its exposition strengthens this guiding thread. Rhythm is also an important element: the magician creates different moments of attention and physical areas of interest to support the effect.[38] But apart from the precise requirements for spatial control of attention, various strategic resources are available to the magician to maintain temporary control and a continuous direction of attention.

Among these strategic resources, a distinction must be made between personal resources, narrative resources, and stage resources. Personal resources are those directly provided by the figure of the magician and are therefore dependent on the magician's experience, skills, and training. These resources include the magician's personality, appearance, clothes, speech, and demeanor—in short, and foremost, everything related to nonverbal communication. Ascanio synthesized the importance of nonverbal communication with the expression "being, presence, and appearance." Narrative resources are key to creating expectations, such as suspense, which, as Alfred Hitchcock said, is a very effective way to keep the audience's attention.* Besides suspense, other narrative maneuvers include stoking curiosity or a sense of the trick's difficulty, planting seeds of uncertainty about success (as when the tightrope walker slips), or using an apparent or feigned failure to lead the audience to expect the trick's eventual failure. Finally, important stage resources for maintaining attention control include lighting and, in certain tricks, music.

In short, synchronizing collective attention and achieving its continuous control over a long period are key to avoiding undesirable contrasts. But doing so requires training and involves, in a unique and characteristic feature of magic, a great deal of effort throughout the presentation from both the magician and the audience. As a member of the audience, you have to follow the presentation of the magic act thoroughly, from beginning to end. Magic therefore makes great demands on viewers from an

* Alfred Hitchcock's elevator conversation is an example of how to apply attention-grabbing methods to an everyday situation. Peter Bogdanovich tells of a time when he was with the brilliant British director in an elevator, the doors opened at a floor, and a group of people entered. Hitchcock took advantage of the moment to suddenly change the conversation: "Blood had splashed on the walls, emptied onto the floor, it kept pouring out of his mouth and nose." He continued to give macabre details and finally commented: "So I had to hold his head and ask him what had happened to it." Just then the elevator reached the ground floor and the doors opened, but no one moved. Finally, the first one to get out had to be Hitchcock. Bogdanovich caught up with him right away and asked, "What did he say then?" "What did who say?" asked the director of *Psycho*. "The one with the blood," Bogdanovich answered. See chrisj01, "Hitchcock's Elevator Story."

attentional point of view, and for this reason, magicians have become true experts in the collective control of attention.

The attentional demands placed on the audience are directly dependent on short-term memory, which is very limited. As a result, the formation of memories (the passage from short- to long-term memory) is subjected to great stress. Further, the demand for continuous attention, added to the excessive information generally provided during the performance of several tricks, can easily lead to exhaustion, saturation, or tolerance, and tolerance always leads to the deterioration of the ability to surprise—as the legendary French magician Robert-Houdin warned in 1868—and even affects the ability to perceive a scene correctly.[39] This is often the case with very long magic shows, which run the risk of being less satisfying. The Argentinean magician Henry Evans, in an interview in 2017 for *Pastomagic,* described the structure of his show: "I like to put a short musical piece in the middle of a long routine, let the audience rest, transmit another emotion, relax, and then I can continue *in crescendo*."[40]

The audience needs time to relax its attention because the cortex expends over 45 percent of all the energy that the brain consumes. One of the fundamental roles of the cortex is information monitoring and attention management—precisely the two most demanding audience tasks in a magic show. The higher the demand, the greater the stress on the audience's brain energy budget, at the risk of a weaker experience or a distorted memory of the show.

Music as a Tool to Transmit Emotions and Synchronize Attention

As we know from horror and suspense films, music can convey emotion as well as, or better than, nonverbal communication. If the soundtrack is removed, most of a movie's emotion is lost; by contrast, eliminating dialogue but keeping the music, as in old silent movies, largely preserves the emotional content of the film. Music is used to create or intensify feelings, particularly through pitch (a single high note can convey excitement, while a low note communicates sadness), tempo (fast tempos usually seem happy, and slow tempos seem sad), and intensity.

The use of music in magic is also powerful, for several reasons. Most important, music can synchronize the audience's attention and highlight the final climax, as well as the intermediate outcomes in acts composed of several tricks. The same image or scene can be perceived differently by the audience depending on the accompanying music.

However, not all music works effectively. Just as music can be a powerful tool to accompany certain magic effects, a wrong selection or poor management can turn it into a harmful adjunct. Sung pieces, for example, are considered riskier and less advisable than instrumental ones because singing is more distracting, and the issue of attention in magic is fundamental. The best music consists of selections that harmonize with the development of the trick and are synchronized with the different events of the expository phase. Unknown or customized music is preferable to the use of classic, well-known pieces; songs by the Rolling Stones or Michael Jackson, for instance, are likely to capture the minds of the audience, drawing them away from the act.[41]

Finally, another factor of equal or greater importance is the type of background music played as spectators enter the room, settle into their seats, and wait for the show to begin. If the music played to welcome the audience is, for example, a very popular Shakira song at full volume, the audience's attention is likely to be attentively exhausted by the start of the show.

Deactivation of Attention in Magic

> Instead of diverting attention to another point, one can lower one's energy level and adjust oneself so that the audience does not notice the important object in the act, or the decisive action that creates the illusion.
>
> —Alfred Binet, "La psychologie de la prestidigitation," *Revue des Deux Mondes*

Just as the magician uses multiple strategies to capture the audience's attention (and avoids evoking it when not convenient), there are times when his methods need to partially or completely deactivate spectators' attention. These are known as the "off-beat moments," a term borrowed from the world of music, where it applies to the upbeats that occur between downbeats.

Recent studies have shown that, after an active capture, the audience's attention relaxes and they tend to blink their eyes synchronously. (On average, we blink once every five seconds.) This state produces moments of deactivation that the magician can take advantage of.[42]

Attention may also be partially deactivated in environments, routines, or stories of special beauty. In magic, beauty is a distraction in that it generates a sense of well-being that contributes to lowering people's guard and the general level of attention. Ascanio highlighted its importance and asserted that, through beauty, "perfection can be achieved." Specifically, he stressed that movements in magic that are wide make beauty possible.[43]

The complete deactivation of attention is achieved, for example, when collective laughter is produced by humorous gags; it is also achieved during the applause, including the applause that occurs after a mini-climax during some acts. Complete deactivation allows the magician to perform preparations or manipulations (such as changing decks or loading or unloading items and gadgets) in full view of the audience members without their noticing.

Special mention should be made of Slydini's technique. By means of his body language, approaching and moving away from the spectator, he achieved subtle cycles of tension and relaxation of attention during the presentation of his tricks. This is a good example of how the choreography of how attention is framed, on both the spatial and temporal levels, can resynchronize audience attention.

The "Deconstruction" of a Magic Trick

Let's try to deconstruct a magic trick into the techniques and methods that we have discussed so far—with the caveat that tricks cannot be standardized: magic tricks using different methods and materials can produce the same effects.

The trick we have chosen, the "Homing Card" by the American magician Francis Carlyle, contains many maneuvers related to attention. As

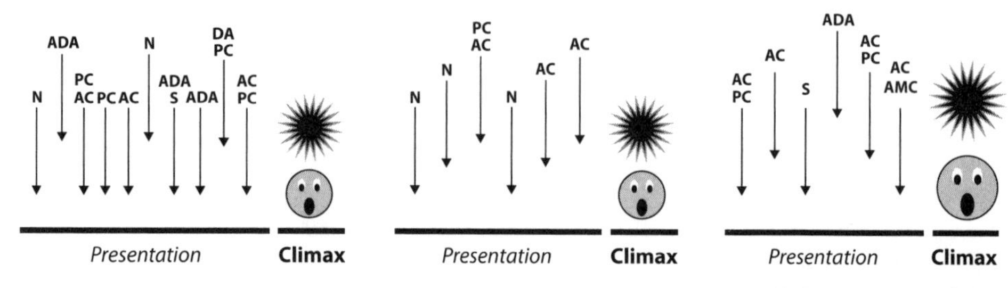

FIGURE 6.5. "Deconstruction" of the effect of the "Homing Card" by Francis Carlyle, according to a version by the Spanish magician Tino Call: The effect develops in three phases, each with its own climax. In the internal life of each phase, various attention control and concealment maneuvers follow one another, but no phase is identical to any other. DA: divided attention; AC: attention capture; ADA: active deviation of attention; N: neutral maneuver; PC: physical concealment (palming, card manipulations); AMC: amodal completion; and S: speed. See the complete trick at:

illustrated in figure 6.5, the trick develops in three phases, each a trick in itself with its own outcome or climax. In the first two phases, a card freely chosen by the audience and then signed travels, surprisingly, from the deck to the magician's pocket. In the third phase, every card in the deck travels to the magician's pocket—except the chosen one, which is the only one remaining in the magician's hand.

This effect is mainly (but not only) based on successive captures and deviations of the focus of attention, as well as on several physical concealments, including some card palmings that must be well executed to avoid becoming a focus of contrast. In the third phase, when the magician holds only one card facedown in his hand—or, in magic terms, "from the back"—a phenomenon called amodal perception occurs (discussed in chapter 7): the audience implicitly assumes that the magician's hand contains the entire deck until the magician reveals that only the chosen card is in hand and the entire deck has traveled to a pocket (see the second part of the accompanying video).

That we see the deck and not the card shows that our perception of a scene does not depend so much on the sensory information received as on our already created expectations about it—a subject to which we turn next.

CHAPTER 7

Perceiving Is a Creative Act, but Everything Is Already in Your Brain

> The world was so recent that many things didn't have a name, and in order to refer to them you had to point.
>
> —Gabriel García Márquez, *One Hundred Years of Solitude* (1967)

To Perceive Is to Interpret

When the potato first arrived in Europe, the French called it *pomme de terre* (earth apple) because its texture reminded them of the *pomme* (apple) that they used so much in their cooking. To perceive is thus to interpret what is being attended to in view of our knowledge about the world. Once the information contained in a visual scene has been filtered, after discarding an infinite number of details, the brain is ready to "label," to understand the information received. Claude Shannon's classic mathematical theory of information has proved useful when assessing the amount of information present in a scene, but in an attempt to go beyond this theory, we consider the more ecological theory of information first proposed by the psychologist James J. Gibson. Gibson coined the term "affordance" to describe the capacity inherent in everything in our environment, natural or artificial, from inanimate objects to even other persons, to suggest meanings and opportunities for actions according to the specific context, thus guiding human behavior in each specific situation.

Magicians' gestures and maneuvers, the objects and gimmicks they use, and the context of the magical effect, including its story line, all work because they afford meaningful perceptions to the audience. Even magicians themselves, as masters of deception, carry very conspicuous affordances to the audience. An affordance is thus always relational; it belongs to the interaction between each perceiving agent with the objects they perceive in the world. Hence, affordances of the same object or person can and do vary dramatically across subjects. The magician's job is to control perception to induce certain affordances that direct the audience to the desired impossible outcome. From this perspective, the cognitive foundation for magic must be found in ecologically available information in the environment, but not only in the way it relates to individual internal perceptions. For most of us, an open hand with the back facing forward affords an empty hand. For a magician, however, or for a spectator who knows the way many tricks are done, it holds the secret behind the effect: an opportunity to hide and transfer a coin or card without being detected.

When the brain perceives, it generates a prediction of what is being observed by inferring and anticipating reality based on past experiences and knowledge about the physical properties of the surrounding objects. The brain carries out the predictive process constantly, millisecond after millisecond, by systematically adjusting and comparing every prediction according to information received through our senses. The brain is particularly well designed to fulfill this function. Two types of connections coexist in the sensory pathways, at every level of the perceptual hierarchy. The incoming sensory information is carried by one type of connection, the so-called feed-forward projections, and the other type, the feedback projections, generate the predictions about the external causes of the incoming sensory input. Although the specific aspects of how these two types of projections interact are still up for debate, this sort of circuit design has found support, for example, in the functional organization of the visual pathway of the brain.

It is worth noting that, by comparing sensory input with prediction, these brain circuits compute how far off the predictions are from the real sensory input—that is, the error we make at predicting the world. As we incorporate already acquired experience into this equation, potential errors that arise in the process of comparing and adjusting are reduced to a minimum. Hence, coincidences make us feel comfortable, while unexpected phenomena, like magic tricks, surprise us a lot.

This process for making inferences is based on hypothesis evaluation. In neuroscience, generally accepted opinion holds that this mechanism follows the Bayesian method—using statistical reasoning retrospectively to infer what current observations may correspond to. In other words, the brain's "best guess" is calculated to interpret what is being perceived based on information obtained from previous experiences. To paraphrase the neuroscientist Dale Purves, "we perceive as we found it useful to perceive in the past."[1] From the catalog of options and variables stored in its memories, the brain manages finally to choose the most statistically probable option.

What emerges consciously is the final choice after the brain has made thousands of ultrafast and unconscious calculations. We are awake yet unaware that these processes are constantly under way. Yet this is how the brain gives perception, as it does memories and decisions, personal touches, or highly subjective components.

Perception, in short, is an essentially predictive process, one in which our brains simultaneously integrate various sources of information, such as immediate sensory information from a visual scene and information from our past experiences in similar situations stored in memory.[2] But why exactly does the brain engage in this predictive process? We know that, among other reasons, the brain works in this way to understand the information projected two-dimensionally on the retina, to minimize the effect of information bottlenecks, and to overcome its slowness. Let's take a closer look at each of these phenomena.

The So-Called Inverse Problem of Vision

Contrary to what you might think, visual information is reflected in the flat surface of the retina as a two-dimensional projection. Here a problem arises, because such a projection has infinite solutions in three dimensions. That is, an infinite number of real three-dimensional objects could generate the same two-dimensional projection depending on the angle of vision, the distance from which the object is observed, and so on. So how do we know what we are looking at? This phenomenon is called the inverse problem of vision.

From the time we were children, we have been creating a catalog, a visual encyclopedia where we store in the form of concepts and implicit memories the existing relations between different projections on the retina and the real objects in the space that produce them. In this "personal catalog," we also keep the name and ecological purpose of everything that surrounds us. We create for ourselves, so to speak, a "catalog for life," and we build it from a very young age. For example, a child playing with a pedal car or shaking a half-full bottle of water and then throwing it is unconsciously taking spatial measurements with the pedal car and weight and volume measurements with that bottle. All of this information will help the child understand what they are seeing when they find themselves with a bottle again, whether or not it is similar to the previous one. Likewise, they will better understand the dimensions of the space they have explored with their pedal car.

Our personal catalog is constantly being updated with the changes taking place in our environment. For corroboration of this fact we need only think, for example, of how our visual idea of a telephone has changed from the time of Alexander Graham Bell to the modern arrival of smartphones. Importantly, at any point our interpretations of what we see and the stimuli we receive are more closely related to the content of our personal catalog—our intimate memory—than to the information we receive directly from the world.

Thanks to predictive processing, the human brain is capable of interpreting a wide variety of signals, from those expressed via nonverbal communication to those expressed in written or oral language. For instance, at this very moment, as a reader of this text, surely you are continually using the word you are reading plus the ones you just read to make a reasonable assumption about the word that will follow. This facility in identifying word patterns explains why you can read so fluently. Identifying patterns is one of the most useful skills we have developed as a species.

Aoccdrnig to a rscheearch at Cmabrigde Uinervtisy, it deosn't mttaer in waht oredr the ltteers in a wrod are, the olny iprmoetnt tihng is taht the frist and lsat ltteer be at the rghit pclae. The rset can be a toatl mses and you can sitll raed it wouthit porbelm. Tihs is bcuseae the huamn mnid deos not raed ervey lteter by istlef, but the wrod as a wlohe.

FIGURE 7.1. Text in which the letters in the words are scrambled except for the first and last letters is nevertheless readable because of our ability to recognize patterns. See http://www.mrc-cbu.cam.ac.uk/people/matt.davis/cmabridge/.

Let's take another example. Read the text in the box in figure 7.1 and see how easily you can deduce the meaning of the words from their context.

Likewise, being able to see a speaker's face while they are announcing something often allows us to predict what they will say. Similarly, we are very often able to start answering a question we have not fully heard or that has not been fully formulated—though misunderstandings may follow. We also tend to use shortcuts when reading, such as when we receive a letter that begins with "We regret to inform you that . . . ," a phrase that immediately makes us imagine a negative outcome when it does not necessarily have to be so.

All of these behaviors follow a pattern: we dispense with connections or filler words that are not crucial to the meaning of certain phrases. However, we can also make mistakes by failing to notice repeated words or by overlooking errors, such as the one in this question: "How many animals of each species did Moses include in the Ark?"[3]

In relying on our own personal catalogs, the interpretation of what we see depends on other individual or collective factors such as age, culture, expectations, and, of course, our own personal experience. For example, a child and an adult will interpret sketch 1 in figure 7.2 differently: the child might see a swordsman or an orchestra conductor, while

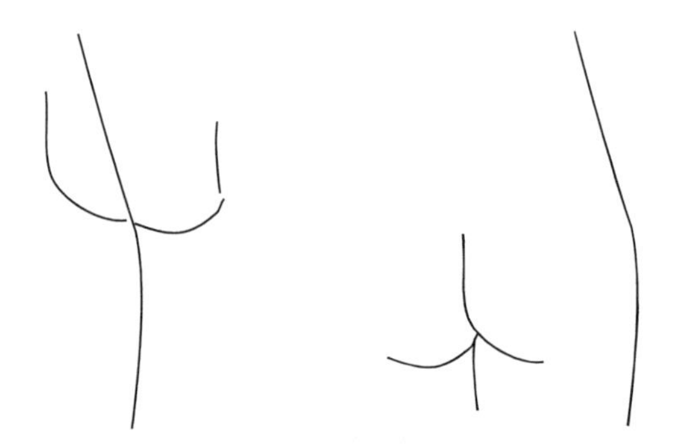

FIGURE 7.2. Two freely adapted sketches of an original by Pablo Picasso: Although the two sketches contain the same lines, their interpretation is very different. Sketch 2 clearly suggests the backside of a nude woman, while sketch 1 is interpreted differently by different people.

the adult is more likely to see a cactus or a trident. Children have not yet fully developed their catalog, their internal model of the world, and so their predictions are much more diverse. Children are also more open to exploring different concepts than adults, who often exploit their catalog to make more conservative predictions.

Bottlenecks in Brain Processing

A second reason why the brain engages in predictive processing is to overcome the unwanted effects of information bottlenecks. In chapter 5, we saw how the strategy for processing information in a relative and not an absolute way, of detecting contrast in both space and time, is used to compensate for the loss of sharpness associated with compression and decompression on both sides of a bottleneck. These bottlenecks also explain the inherently interpretative strategy of the phenomenon of perception, because information is always lost in the process.

By way of comparison, let's take a look at what differentiates the brain from a digital image processing system. In 2015 alone, pictures taken in digital format outnumbered the total number of pictures taken in the entire history of analog photography.[4] Since then, the output of digital photography has progressed exponentially, creating a problem of the first magnitude, so great that we have not yet been able to gauge all of its consequences. Since the late 1990s, there has been a growing demand for digital cameras with increasingly precise resolution to capture "reality" in the greatest possible detail. But cameras capable of greater resolution inevitably create files that are bigger and harder to share. At this pace, the fiber-optic cables supporting the internet will reach their maximum capacity in a few years. Transmitting information over the internet also consumes a very significant amount of energy: it is estimated that, for example, in the United Kingdom the internet already uses approximately 12 percent of all the energy consumed, equivalent to the production of three nuclear power plants. Furthermore, consumption doubles every four years.[5]

Two types of solutions are proposed to meet this challenge. The first is to expand channel capacity by increasing the number of cables and restructuring the physical capacity of the internet, an expensive and complex undertaking. The second is to reduce the size of the transmitted signal. To this latter end, every digital camera already includes an information compression algorithm that greatly reduces the size of the files it generates to make them more manageable. These compression algorithms are mathematical tools that transform the image into its fundamental components before transmitting that information. In turn, the receiver has the complementary decompression algorithm that recomposes the image with very little loss of information. This is how .jpg and similar files commonly used today work. The high-resolution photos we take are transmitted in low density and reconstructed at the destination, greatly reducing the related energy consumption and optimizing the channel occupation.

The brain's visual system faces a very similar problem: we start from a bottleneck between the high-resolution camera (the eye) and the processing unit (the brain), which cannot possibly receive all of the information that the eye picks up. But in the brain evolution has provided slightly different solutions than we have devised for transmitting digital photos. As we have already explained, the brain relies on compression to receive information from the eye. But this process entails a very significant loss of information, which has consequences for the way the brain interprets an image. Just like a digital photograph, the low-density image that reaches the brain has to be decompressed to recover its initial resolution, and then, to achieve an image of similar size to the original, the lost information must be filled in. Here is the big difference between a digital camera and the brain: in a process of interpretation rather than passive transmission, we fill in the gaps in the images with data inferred from our memories.

The inferences that we make, our interpretations, are nonetheless very precise and useful, being based on the infinite experiences stored in our memories about the physical properties of objects in the world and their relationships. As we have seen, some of these interpretations, which we have been accumulating throughout our lives, are subjective—personal and nontransferable. Others are shared, the product of the culture in which we have grown up, such as Westerners' and East Asians' different aesthetic preferences for forms of representation of three-dimensional space. Still other inferences may be innate and due to the evolutionary course of our species, such as an aversion to spiders or snakes.[6]

The Brain Is Slow

Yet another important limitation of the brain as it is forming predictions is its slowness. The brain is slow: not only do we process just a very small proportion of the constant barrage of sensory stimuli, but we process with a delay of at least a third of a second—or more if the mind is focused on other things.

Under normal conditions, this delay does not have major consequences. It hardly matters to us if the music we are listening to on our player is delayed a third of a second, nor is it crucial if, while we are closely watching the precise moment of a sunset, there is a slight delay before we see the sun disappear.*

* After all, many stars we see in the night sky have already disappeared (though not all of them) because the brightness associated with their final explosion reaches us millions of years later.

One reason for the delay in our brains is that all the information we obtain through the senses is processed by groups of neurons located in different parts of the brain, sometimes separated by relatively long distances. This type of processing requires time. For example, it takes 20- to 40-thousandths of a second for information about a visual scene to reach the first area of the cerebral cortex where it will be processed. The full process of recognizing an object, such as a face, takes between 125- and 200-thousandths of a second, and much more time will pass before that visual information can significantly influence our behavior.

The experimental observation of this delay in some processes, such as certain types of decision-making, and the fact that we are not aware of it—for example, when we make concrete decisions—has led to heated discussions about the existence of free will. Experiments by Benjamin Libet have shown that by the time we become aware that we have made a certain decision (such as pressing a button when a signal appears on the computer screen), the brain actually already made that decision unconsciously at least one-third of a second earlier.* These results pitted those convinced of our capacity for free will against those who would argue—based on these and other experiments—that free will is an illusion.

To counteract its slow speed, the brain has needed to learn to make very accurate inferences. If we are on average one-third of a second behind, it becomes very difficult, with no predictive power, to imagine how we are able, for example, to drive cars at high speed, play tennis, or perform most everyday activities, such as crossing the street or avoiding obstacles as we walk. Does anyone think that, to defend against a penalty kick, goalkeepers throw themselves to the side where they "saw" the player kick the ball? Actually, goalkeepers almost always make a certain leap in the hope that they will be lucky enough to have chosen the correct side. (Of course, to increase their chances of success, some top goalkeepers learn the "habits" of their opponents.) A study

* Libet, "Unconscious Cerebral Initiative." More recently, these experiments have been reproduced using functional magnetic resonance techniques, and it has been possible to predict the unconscious moment of decision-making about ten seconds before this decision reaches the conscious plane, as explained in chapter 5. See Soon et al., "Unconscious Determinants of Free Decisions."

conducted at the University of Amsterdam analyzed the penalty kicks taken in all of the World Cups between 1982 and 2010 and concluded that when a team is winning or tying, the goalkeeper attacks 50 percent of the time on the right and 50 percent of the time on the left. But if a team is losing, the goalkeeper has a preference for the right side (seven times out of ten), perhaps because most goalkeepers are right-handed and, when they are under a lot of pressure, they tend to go to their natural side, the right side.[7]

An example from tennis is even more illuminating: on a court approximately 24 meters long, when a player serves the ball at about 200 kilometers per hour, the opponent will have only about 400-thousandths of a second to detect its trajectory, direct their body and the racket toward it, and finally hit the ball in the right direction to return it to the opposite court. Using visual information—including details of the opponent's movements moments before the serve and the choreography of the opponent's movements—as well as their own previous experience, the player's gigantic brain database is capable of providing them with expectations about where and how the ball will come.

The brain's normal processing delay is compounded by other internal maladjustments, such as the different processing speeds of sensory stimuli. Auditory information is transmitted through the brain more quickly than visual information. It is for this reason, for example, that sprinters leave the starting position more quickly when they hear a detonation to start the race than when they see a flash. In coin tricks, the sound of coins striking one another and the sight of several coins being passed from one hand to the other are perceived to be simultaneous if they happen no more than 70 milliseconds apart, allowing the magician to create the illusion of synchronicity.

But just because the brain uses interesting and diverse strategies to compensate for these delays does not mean that it will always be victorious. Let's look at an example. You can grab a subway ticket by one of its corners with your thumb and index finger and invite another person to hold their thumb and index finger beneath the ticket, ready to capture it at the moment it falls. If you do not warn the person and release the card quietly, no matter how close the other person's hand is, they will almost never succeed in catching the ticket because the visual information is received too slowly and their motor reaction will consequently be delayed. If, on the other hand, you also make a sound when you release the ticket, the other person is much more likely to succeed in grabbing it.

Human Beings Anticipate the Future

As we have explained in this chapter, the brain relies on predictive processes to compensate for its functional limitations. In this way, most sensory and motor systems have learned mechanisms of anticipation that is, capacities stored in implicit memories that allow us to react seemingly immediately to predictable events. This capacity for anticipation, this way of thinking about the future, is a characteristic feature of adult human beings. Despite sharing with other species a myriad of conditioned responses (for instance, "If I see that kind of animal I must flee immediately, as it might finish me off"), humans are capable of continuously and unconsciously making predictions about the immediate future. (We also have the ability to make complex longer-term predictions, although the processes required to do so are quite different.)

This ability to imagine a series of immediate events that have not yet occurred is made possible by, among other areas, the frontal cortex of the human brain, which is precisely the last part of the brain to have evolved, the part that matures most slowly during development, and the part that is first to deteriorate with age. It is the frontal cortex that enables us to think about existence in the long term, to experience the future before it happens, and to imagine objects and evens that do not exist in reality—in short, to daydream. In contrast, some people with prefrontal injuries lose the ability to plan and no longer experience surprise in the face of the unexpected or anguish in the face of the unknown.

It is extraordinary how the brain has perfected its ability to anticipate events, especially in dangerous and unexpected situations. For example, when you are about to cross a busy street, your brain makes thousands of calculations and even imagines that oncoming cars are traveling at faster speeds, exaggerating your perception of risk and giving you more time to decide whether to cross.[8] It takes at least 200 milliseconds to bring information from the eyes to the inferior temporal cortex of the brain, which is where objects are identified, so your visual system anticipates the cars' speed as being at least twice as fast as their real speed to give you time to act accordingly and avoid getting hit.

Another example: In countries where people drive on the right side of the road, you instinctively look to the left before crossing a street, because the brain forecasts that cars will come from your left. Thus, when you travel to London, even though you know that cars are coming from the opposite side from the one you are used to (and despite the many reminders in pedestrian crossings), you are continually frightened. For this same reason, when we watch a thriller, we cannot help but speculate about its outcome, and knowing this, filmmakers make us suffer by constantly playing with our expectations. As Alfred Hitchcock, the "Master of Suspense," said, "There's no terror in the bang, only in the anticipation of it."

Be that as it may, because humans are best adapted to what is predictable, coincidences will always give us a sense of satisfaction. This is why, when faced with coincidental phenomena, we often cannot resist the temptation to grant them causality. The fact is that inferences are the basis of daily functioning, in which our experience plays a fundamental role. This is also the reason why so many people have an aversion to the new or unknown, and even why any new experience or travel to an unknown place is very hard for some people. This is not to say that uncertainty is always detrimental. A certain degree of surprise is always necessary to update our internal model of the world (our predictions), and that is why we like magic shows or thrillers in which the plot changes suddenly and unpredictably. But everything has a limit, and living in a world dominated by uncertainty would be untenable, since it would be very difficult to interpret or interact with it effectively.

Moreover, we often cannot adequately perceive the beginning of an unforeseen event, and that can lead to various consequences, such as not knowing how to react appropriately. Magicians, of course, rely on the power of the unexpected event, taking great advantage of the fact that the audience cannot predict what will happen in their show.

Through their actions, magicians generate in the audience a particular and private environment and provide a unique set of affordances that, during the trick, only they control, leading the spectators to anticipate a solution that, in the end, will be frustrated by an ending that seems impossible.

Magic as the Art of the Unexpected

> Just so the conjurer never reveals in advance the full nature of a trick, that the spectator may not know where to center the attention.
>
> —Max Dessoir, "The Psychology of Legerdemain" (1893)

The natural and unconscious capacity of human beings to anticipate most events of daily life makes possible wonderful effects in magic. Many magic tricks surprise us with their seemingly impossible, totally unexpected outcomes, illustrating how magic takes advantage of the brain's strategies. In other words, magicians anticipate the expectations of spectators, then contradict those expectations. Together with the unexpected, these contradictions achieve the illusion of impossibility.

A great precept in magic is to circumvent the brain's predictive ability by never revealing in advance the nature of the effect being presented. (Not all magic tricks do so, however, since the "supposed" outcome of some tricks should be announced.) Many magic tricks base their strategy precisely on the fact that the audience is not in a position to anticipate which effect or effects will be produced, and therefore continuous control of the audience's attention can be left in the hands of the magician. This strategy works because when we observe something new or unexpected for the first time, we always have to make a great effort to stay focused.

On the other hand, if spectators know what is going to happen, they do not have to make the same effort, especially if the same thing happens repeatedly; in this case, spectators will notice other details and the magic trick's effectiveness will be ruined. Reduced effort to focus attention leads to greater attentional flow facilitating the capacity to process.[9] Thus, in magic the "principle of non-anticipation" is essentially a matter of attention.

Another precept in magic is the "principle of nonrepetition," according to which one should avoid presenting tricks that use the same effects or methods again and again in the same routine, or the same method several times in a single trick. Humans tend to believe that things that are repeated over and over again have the same cause, so a continuous repetition of methods encourages the audience to test and then debunk any hypothesis of perception previously put forward. Thus, the second time you see exactly the same thing, you are already in a better position to anticipate what is going to happen, having quickly learned where to direct your attention instead of letting your attention be directed. Experimental laboratory studies have confirmed that repetition extinguishes the ability to manipulate and that, in general, prior information about the magic act to be performed significantly increases the ability of participants to detect the method.[10]

Occasionally, the effectiveness of a magic trick is maintained in spite of repetition throughout the act, as has been demonstrated with the false transfer of a coin from one hand to the other.[11] Usually, however, the coin transfer appears to be happening because the magician continually changes the method to produce the same or a similar effect. In fact, this is precisely how the young Canadian magician Dai Vernon fooled the legendary Hungarian American magician Harry Houdini with a personal variation of the "ambitious card trick"; in this classical effect, described by the French magician Gustav Alberti in 1877, a card, after being lost in the deck, always reappears at the top.

Developing Hypotheses Automatically: Amodal Perception in Magic

Amodal perception (sometimes called "amodal completion") is the tendency to automatically and immediately complete any object that we see partially hidden. Even when information is scarce, the visual system completes the parts of the objects that are not visible to us, giving those objects a very concrete perceptual form and meaning. As we will see, it is very difficult to overcome this unconscious process, in part because

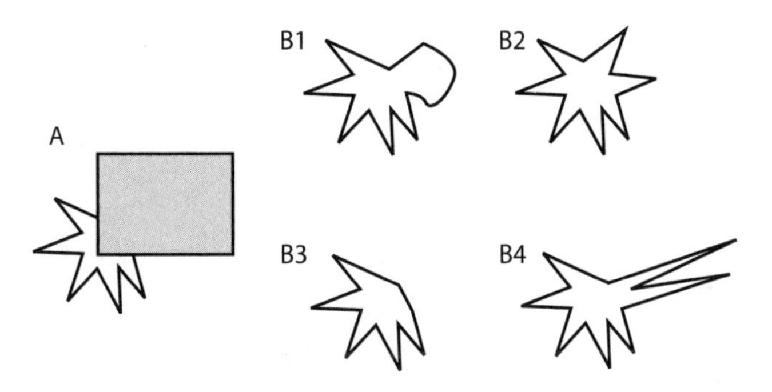

FIGURE 7.3. An example of amodal perception: The viewer imagines that A is like B2 and does not realize that it could be any other B.

the way in which these images are completed is, again, a statistical prediction based on the experiences we keep in our memory.

Let's look at the details: The mental process of completing an object with its invisible parts or grouping together parts of a visual scene follows certain rules. Known as Gestalt laws, these rules describe the constancy or regularity with which we group objects and organize them, as well as how we interpret their continuity when they are partially covered.* Specifically, our visual system always applies the simplest, most regular, and most symmetrical interpretation of an object or scene that is partially hidden (figure 7.3).

Many magic tricks take advantage of these rules to achieve the illusion of continuity in the spatial domain. For example, tricks using partial concealment or screens take advantage of the Gestalt law of continuity, which states that "elements of objects tend to be grouped together, and therefore integrated into perceptual wholes if they are aligned within an object. In cases where there is an intersection between objects, individuals tend to perceive the two objects as two single uninterrupted entities" (see figure 7.4).[12]

The main applications of these apparent continuities can be found in rope and string tricks,[13] in bent spoons, or in Chinese linking rings, as

* Gestalt laws were formulated within the Gestalt movement that originated in Austria and Germany and focused on studying the psychology of form and spatial configuration.

FIGURE 7.4. An illustration of the Gestalt law of continuity: The knife appears to cut the arm of the magician David Valencia.

well as in card tricks, such as the torn-and-restored card trick (see the Guy Hollingworth version of David Regal's "Piece by Piece"), or in linking cards (see Paul Harris's "The Immaculate Connection"). In all these cases, the magician's hand acts as a screen or partially covers the objects at specific moments and under conditions such that the audience's interpretation of what is hidden is counter to reality. In the Chinese linking rings trick, the magician seems to link and untie the rings at will, apparently defying their metallic solidity, but the magician is showing the ring while covering the gap, so we automatically perceive the whole ring. Likewise, we do not question the nature of what lies behind an occluder. It escapes our conscious control because amodal perception (the process of completion) acts by default, never fails to act, and is immediate.

Interestingly, the same effect can be repeated over and over again. A very striking case is that of the magic effect based on manipulating several balls in one hand, making them appear and disappear continuously: the illusory experience persists even after the audience learns that the balls are shells, not solid balls. As it happens, the three-dimensionality of the visible surface of the shells is sufficient for them to be perceived as a solid sphere.[14] What happens is that our perception of the hollow shell cedes to the more common and simpler perception of the sphere; after all, the two objects produce the same retinal image in the spectator. We are left with this solution and consider no alternative. It seems that one of the assumptions most relied on by the audience in a magic show is symmetry, both in static situations and in action sequences.[15]

But the most interesting thing about amodal perception is that it stands up well to repetition, unlike the capture and active diversion of attention, whose effectiveness declines very quickly if repeated.[16] This is the case because, in magic, concealment based on amodal perception has the advantage of invoking automatic assumptions that are not suspicious and do not induce viewers to "rewind" or think about what is being hidden from them. In short, such forms of concealment do not produce undesirable contrasts.

Vebjørn Ekroll's group argues that disproportionate weight has been given to misdirection and attentional control mechanisms; they attach greater importance to these effects of amodal perception as being "cognitively impenetrable perceptual mechanisms."[17] In other words, these effects should play a central role in many magic tricks because they do not register consciously.

Finally, opposite amodal perception is a perceptual illusion of "amodal absence": that which is not seen is considered not to exist. In truth, it is not exactly the reciprocal of amodal perception. The phenomenon is best described as the inability of the audience to perceive a magician's affordance, such as the magician's capacity to hide any object in the palm of their open hand; the illusion of empty space surrounding people or objects that are levitating; or even the illusion of levitation around empty space, beyond which observers imagine that something supports the object or person apparently floating. Like amodal perception, this inability to perceive an affordance is considered to be an automatic, unconscious perceptual illusion.[18]

In short, we perceive the world based on our previous experiences. The present is always captured and shaped from the past. And as we will see in the next chapter, our memories, whether shared or deeply personal, are fundamental to this process.

CHAPTER 8

To Remember Is to Rebuild

The Function of Memories

We think of memory as the seat of our recollections and learning, but our brains use different forms of memory to manage information and understand reality: to think, decide, plan, and project ourselves. In short, memories play a part in all of our cognitive and emotional processes.

A specific memory corresponds to the activation of neurons functionally interconnected and synchronized with each other. Today we know that the neuronal interrelationship that defines a memory is a highly dynamic and versatile process and that different types of memories, such as sensory memories, short-term memories, and long-term memories, are defined by the strength of their connections.

To this point, we have discussed briefly the importance of memories to how we perceive and decide, whether we are witnessing a magic trick or simply moving through the world. In this chapter, we immerse ourselves in the universe of memory: both explicit, or declarative, memories, which are long-term memories of a conscious nature; and implicit memories, the other great type of long-term memories, which include motor or procedural memory and perceptual or emotional memory.

We start with explicit memories, which are consciously expressed and can be acquired in one or a few repetitions. Explicit memories can be divided into two types. Semantic memories, the first type, are of concepts and names; they involve areas of the cerebral cortex that specialize in processing faces, objects, actions, language, and many other categories of knowledge. The second type, episodic memories, collect events, occurrences, and experiences and are formed in the hippocampus. They are commonly thought of as recollections or deliberate memories about the world and our personal experiences.

In the next section, we begin with explicit memories—with an emphasis on episodic memories because they are more relevant in magic—and examine how they are formed, forgotten, and evoked. We will see that to memorize and remember, we use strategies similar to those used when we perceive and that magicians have learned to interfere with these strategies. The world of magic has not only incorporated the neuroscience of memory into the performance of magic tricks but has also taken advantage of it to better cover its own secrets.

Explicit (Declarative) Memories

> We do not remember days, we remember moments. The richness of life lies in memories we have forgotten.
>
> —Cesare Pavese, *This Business of Living*

If you were asked to describe everything you saw yesterday in all its detail, could you do it? Even if you could, you would need all day to retell all of it. A great peculiarity of the construction of episodic memories is that only essential features are kept. We keep only some details because, even if we could store all of the information we receive, our brains do not have the power to process it.

The brain stores information in parallel, rather than sequentially, for reasons of efficiency. Storing memories sequentially, like a recording made on a videotape, would require inconceivable brain size. What does seem clear is that details of past experiences are forgotten. A recent study by the neuroscientist Gabriel Kreiman and his group, done outside the laboratory under natural conditions, quantified the ability to remember details over the course of one hour in real life. Kreiman's group videotaped and tracked the eye movements of their subjects as they walked along specific routes, then assessed whether the subjects could distinguish the video of their own episodic experiences from the videos of other participants who followed the same route on different days. The researchers found that participants' chance of remembering the details of the scenes they had walked through were roughly fifty-fifty.[1] Unable to store every detail, the brain stores only the traits or concepts that will help it to form categories and create relationships to establish various sets of knowledge.

When storing memories, we strip our perceptual experiences from all contextual or particular information down to their very bones, to the point that we retain mostly stereotyped concepts, events, and relations. For example, we memorize the concept of a "cat" (based perhaps on our own cat), but we cannot save the details of every cat in the world. Over time we link the mental representation of "cat" to direct categories, such as "pets" and "animals," or to related categories, such as "hair," "company," or "house." This associative versatility of storage affords the flexibility necessary to cover many concepts with few resources.

Episodic memories are formed in the hippocampus, a part of the brain located in the temporal lobe. This very interesting brain area intervenes in other important functions as well, such as spatial navigation and conflict processing, among others. The cognitive neuroscientist Rodrigo Quian Quiroga discovered conceptual neurons in the human hippocampus. These are neurons that encode associations and are specifically activated when we are conscious of seeing or hearing a thing, place, or person. According to Quian Quiroga, these neurons seem to be specific to the human species and they form the basis of our memories.[2]

When information passes through the hippocampus, it is classified into several categories. Instead of storing all of the information in a single area of the brain functioning like a computer's hard disk, the hippocampus redirects the memory fragments to different cortices (visual, olfactory, gustatory, and so on) and to other areas of the brain. Thus, for example, words are stored in the temporal lobe, colors and other visual information in the occipital lobe, and the sense of touch and movement in the parietal lobe. Emotional memories, as we will see in the next chapter, are processed in the amygdala. When a memory is evoked, it is reconstructed in the hippocampus. Thus, the associative capacity of this great assembling organ plays a key part in memory reconstruction.

Stages of Long-Term Memory Formation

In the formation of long-term memories, including explicit memories, three stages can be distinguished: encoding (acquisition), consolidation (storage), and evocation (retrieval).

Encoding is the process of acquiring information that reaches the brain's hippocampus, resulting in the creation of networks of interacting neurons. The intensity and selection of this information depend on attention: the more attention you pay to a piece of information, the more likely it is to be encoded into your brain as a memory. However, the process is a fragile one, and magicians have learned to manipulate it through all kinds of distractions. If we are distracted while performing a task, we stop paying attention to it, and later we will have trouble remembering it—we will not be able to find the keys or the assignment, or we will miss the appointment.

Once encoded, memories are consolidated, or maintained, when they pass from short-term memory to long-term memory through a process of learning by repetition. Sleep—specifically non-REM sleep—is a vital stage in this process. In this respect, a phrase often repeated by one of our high school teachers—"Lesson slept on, lesson known!"—remains a good example of scientifically sound popular wisdom.

The consolidation of memories and long-term learning involves protein synthesis and structural remodeling in dendritic spines, which promote synaptic plasticity. The duration of episodic memories of events, occurrences, and experiences depends on whether they are repeatedly evoked; otherwise, the neural connections lose stability over time and the memories fade.[3] But not all types of memories are consolidated in the same way. Semantic memories of concepts and names tend to be more stable than episodic memories. Amnesic patients, for example, retain verbal and mathematical skills, as well as other functions required to evoke these memories.

Some memories are very durable, depending on where they are stored. Our recall of language and music, for example, is preserved in the temporal lobe and in a subsystem of long-term memory located in the frontal cortex and thus tends to remain very strong. The loss of neurons caused by the degeneration that occurs in some pathological situations, such as Alzheimer's disease, affects first the hippocampus and then the entorhinal cortex, key areas for the formation and consolidation of new episodic memories. Even so, Alzheimer's patients with musical abilities retain them for a long time because these storage processes remain relatively intact until the advanced stages of the disease.[4]

Even healthy individuals lose faculties as they age, resulting in amnesias of episodic memories. Our learning abilities are affected when a weakening in synaptic connections and in several neurotransmission systems impairs neural modulation. These changes occur in areas of the frontal cortex (involving executive functions) as well as in areas of the medial temporal lobe (affecting the storage of new memories).

Memories Recorded in Especially Emotional Circumstances

Though many long-term memories require repetition in order to be consolidated, some very vivid episodic memories, formed in association with a strong emotional situation, are instantly recorded and never forgotten. These are known as "flashbulb memories," an analogy to the brilliance of a photographic flash. These memories may be autobiographical, featuring moments of particular personal significance or trauma, or they may be collective memories of historical events that are remembered in a special way. Each individual carries their own list: the day in 1969 when man first stepped on the moon; a major assassination, such as that of President Kennedy in 1963; the terrorist attacks on the Twin Towers in New York City in 2001; or the 2004 attach on the Atocha metro station in Madrid.

What is interesting is that though these emotional memories are easily remembered, the details are just as imprecise as those of non-emotional memories. The reason is the same: only a small part of the information flood that passes through the brain is collected. Even though these events are difficult to forget, and even though people are very confident in what they remember, the memory is very partial, or unreliable—just a few features or concepts are retained.

Elizabeth Phelps, a leading neuroscientist known for her research on emotional memories, conducted studies with witnesses of the 9/11 attacks in New York City. Using magnetic resonance imaging, she demonstrated that years later, when remembering the attacks, people who lived downtown near the Twin Towers still experienced an activation of their amygdala (the area of the brain where emotional memories are processed); no such activation happened with people who lived farther away.[5] This selective activation of the amygdala demonstrates the great emotional component of the lived experience, with repercussions for the processes of encoding, storage, and retrieval of associated memories. People who lived near the Towers at the time of the attacks had very vivid memories; however, they did not record peripheral or neutral details, such as how they were dressed. Phelps prepared a questionnaire that she presented to several volunteers within a week of the attacks, asking where they were when the planes hit the Twin Towers, whom they were with, what they were doing, how they felt, and so on. She posed the same questions to the same respondents later, at one, two, and ten years after the attacks. At each contact, she was surprised to find that some people had changed their answers. Participants who at first claimed that they had not witnessed the attacks because they were on a business trip later vehemently claimed to have witnessed them in situ, and even to have seen people trapped on the upper floors throw themselves into the void (events widely publicized on television). Even more surprisingly, when the researchers showed the participants their original responses to the questionnaire from the week of the attacks, practically all of them persisted in confirming the truth of their current response, countering that at the time of the attacks they must have been in shock and surely did not know what they were saying. Phelps's study clear demonstrates that memories of episodes with a strong emotional component are not reliably stored and are reconstructed later, at the time of their retrieval.

Phelps also observed that although the participants' confidence in their memories was consistently high, the memories themselves began to deteriorate after a year. She discovered that, after ten years, although inconsistencies remained, inaccurate details in participants' episodic memories had been corrected under the enormous influence of the media, films, documentaries, and books about the events.[6] Clearly, external influences are able not only to create erroneous memories but also to correct them.

Studies have also tried to discern whether there are elements of events with great emotional impact that are more likely to be stored in memory than others. Sometimes it is neither the emotional intensity nor the duration of an experience but the sequence of events that it comprises that determines what and how it will be remembered. Kahneman and his group conducted studies to evaluate the painful memories of patients who had to undergo an uncomfortable colonoscopy (at a time when this procedure was still done in a long and painful way). Researchers carefully recorded the length of time that pain was experienced during the procedure and analyzed its influence on patients' later recall of the experience. Surprisingly, the duration of the procedure (colonoscopies lasted from a few minutes to over an hour!) did not affect the pain rating of the recalled experience. Instead, the worst memories—those rated as more painful—were related to intense peaks of pain, especially if they occurred at the end of the procedure. In one of the studies, researchers left the colonoscope in place for a few minutes longer than required for clinical reasons, but without causing any further discomfort; patients in this group were precisely those who later reported having the least unpleasant overall experience, as if they had forgotten the intense pain they had felt in the initial phases of the procedure.[7]

Other studies have corroborated this same bias, known as the "peak-end rule," which highlights the importance of the final part of an entire experience. For instance, we may have a bad memory of a vacation because of negative events on the last day of our trip, even if the majority of the vacation was pleasant. These are biased memories that may unfairly influence later decisions.

What questions do studies like these raise in regard to the experience of magic? What differences are there between the spectator's experience of magic in the moment and what he later remembers—after witnessing an effect, or several tricks, or a whole show? For some spectators, the experience of a magnificent magic show is unforgettable, to the point where they take magic on as a hobby or even devote themselves to a career in magic. This common phenomenon—the experience of the "unforgettable" in magic—has never been studied in detail. What do spectators remember best, the last effect or a particularly emotional part of the show? If a very striking effect has been produced, what details are remembered? Is there a primary memory or a type of memory that the spectator retains after the show? Does the entire audience have the same memories? Does the spectator who has been invited to participate in a trick have different memories from those of other spectators?

Without a doubt, the answers to these questions, based on real situations rather than laboratory experiments, would surprise many magicians and provide insights that would improve and magnify the experience of the impossible. Further, scientific study of the memorability of magic tricks would also hold valuable lessons for the neuroscientific study of memory, and of cognition in general, in real-world environments. We recently made a first attempt in this direction. Our study showed that the memory of a magic trick decays over time—as do other episodic memories. We also found, however, that the recency effects that render the last tricks performed more memorable when recalled right after the show are no longer present weeks or months later, suggesting that short-term memories were not converted into long-term memories.[8] This preliminary work is, to our knowledge, the first scientific study of the memorability of magic tricks, and it illustrates how useful magic techniques are in studying memory.

We Need to Forget in Order to Remember

As we explained in chapters 2 and 3, brain networks are interactive and very elastic. While new connections store new memories, room is also made for new memories by impairments or disconnections that prompt us to forget. Forgetting, then, is not a mistake but an essential and necessary process that the brain actively promotes. The brain does not rest. It is continuously changing, as remembering and forgetting always entail structural, anatomical transformations.

Sleep is an integral part of this process. The brain we wake up with is not exactly the same one we went to sleep with; during sleep, we consolidate memories, strengthening connections of recent learning, and we eliminate memories so that fresh memories, especially episodic ones, can be added the next day. This cleaning or purifying process is indispensable for normal brain functioning. We see evidence of this in people who experience sleep deprivation, which impairs hippocampus-dependent long-term memories. Sleep-deprived people quickly lose concentration and the ability to acquire new information, among other cognitive losses.

The Reconstructive Character of Memory Evocation

> He was still too young to know that the heart's memory eliminates the bad and magnifies the good, and that thanks to this artifice we manage to endure the burden of the past.
>
> —Gabriel García Márquez, *Love in the Time of Cholera* (1988)

The evocation of explicit memories is usually instantaneous and automatic. Contexts and "hooks" aid recall so that, for instance, you remember a recipe more easily if you are in the kitchen where you usually prepare meals—just as Proust's narrator involuntarily recalls his aunt's home after tasting a madeleine dipped in tea. Being in the place where memories were acquired facilitates recall, as does being with those present when the event in the memory took place. Remembering facts from your childhood, for example, is easier when you are among former classmates than it is among your current coworkers.

The evocation of memories can also be flawed. Sometimes we suffer blockages and our search algorithms fail. Sometimes you have some piece of information "on the tip of your tongue" but cannot recall it. These are temporary failures that are usually resolved after a short time.[9]

These lapses occur because the reconstruction of memories is a process based on associations, which can be as illusory and unreliable as perception. Surpassing by far any existing computer, the brain has a great super-searcher capable of associating concepts in the most extraordinary and immediate way ever known, reconstructing memories and making inferences at great speed and in a highly refined process. When a memory is evoked, the brain organizes and gives coherent meaning to details of the recalled event stored in different places. As we evoke memories, however, we reconstruct them, and the resulting memories are never exactly the same. Our reconstructions omit, simplify, and add. In the evocation of episodic memories, the brain is creative and undiscriminating: if the memories are vague, the brain is capable of inventing or importing details, and when there are memories with gaps, the brain tends to fill them in.

In short, we can say that we never evoke a memory in the same way. For this reason, our autobiographical memory is seasoned with knowledge and experiences contributed by our parents, siblings, relatives, and friends. Inevitably, some details of our autobiographical memory are fiction, including in particular all memories acquired prior to the age of three or four; at those ages, the processes of memorization have not yet matured. Further, our memories of publicly known events may not have been created from our direct experience of them; the brain often perfects or complements memories of public events with information aired later in the media—one reason why there is serious doubt as to whether eyewitness testimony can be sufficient evidence to convict a suspect of a crime.

Memories are not just reconstructions; we are very selective in the details we retain from a certain experience. Several people who underwent the same experience will never remember it in exactly the same way; each person will retain different details. Furthermore, memories are biased by our knowledge, influences, and beliefs. Both collective and individual memories suffer from the same limitations and biases.

In short, we can not compare the accuracy of long-term memory with the fidelity of what is recorded by a video camera. Memory reconstruction should rather be thought of as similar to the work of archaeologists, who, from a few bones, must imagine a whole skeleton, knowing that they may be wrong. So all memories are false to some degree—but of course false memories can also be implanted.

False Memories

> The difference between false memories and true ones is the same as for jewels. It is always the fake ones that look the most real, the most brilliant.
>
> —Salvador Dalí

In special circumstances, memory recall can give rise to false memories, be they distorted memories or memories about situations that never happened. Unfortunately, we are never fully aware of these distortions, and false memories can be vivid and assumed with great confidence, to the point that convincing a person that their memory of a certain event is wrong is very difficult. False memories, in this sense, must be considered failures directly related to the strategies used by the brain to encode and consolidate memories.

As we discussed earlier, distractions can erode potentially memorable information, as can stressful situations or scenes with excessive information. What happens first in such circumstances is that our short-term memory is overloaded, so we remember only a fraction of what occurred. Imagine that a person is a victim of a robbery and is being threatened with a weapon. In this situation of intense stress, with short-term memory saturated, the person's attention will be focused on the weapon and the rest of the scene will recede into the shadows. Only a faint representation of the attacker's face will be encoded in long-term memory. It is not uncommon, moreover, for these highly emotional circumstances to be followed by episodes of amnesia. Although it may seem surprising, memory overload from an extreme situation can also occur when witnessing a magic act.

Disinformation and false memories are also created through alterations in the phases of consolidation or evocation. In the field of memory consolidation, a type of disinformation that is easy to provoke is the "attribution error," an erroneous inference due to apparently logical assumptions. For instance, if you see a person you dislike making a blatant mistake, you are likely to attribute the mistake to the person being bad or careless. If you see somebody whom you appreciate making the same mistake, however, you will tend to attribute it to their circumstances.

Suggestion is another powerful technique by which the reconsolidation of memories can be manipulated, to such an extent that, in propitious circumstances, false memories can be implanted. A very influential article by Elizabeth Loftus, the scientist who has gone most deeply into the formation of false memories, describes an experiment along these lines. Participants in the experiment were asked to remember childhood events that had been recounted to the researchers by a parent, an older sibling, or another close relative. Loftus and her colleagues prepared a booklet for each participant containing one-paragraph stories about three events that had actually happened in the participant's life and one that had not. The false story was a plausible one about a trip to a shopping mall in which the subject got lost at about the age of five. After reading the booklet, almost one-third of the participants remembered either partially or fully the false event constructed for them.[10]

Memory reconsolidation can also be affected by disinformation and "framing," that is, the slight variation in how a question is asked or a fact is presented. Loftus conducted a study more than forty years ago that is still an important reference today. She asked two groups of subjects to watch a video of a traffic accident. She then asked them how fast the cars were going. The story of the accident was not the same in each group: in one group, the cars came in "contact," and in the other group, they "hit," "bumped," "collided," or "smashed" into each other. The participants estimated the speed at which the cars were going in direct relation to the verb used in the story: those who saw the cars as "smashing" estimated that the cars were going faster. A week later, when the same people were asked if they remembered seeing broken glass at the scene of the accident, surprisingly, those who saw the cars "smash" claimed to have seen more glass—although in the video there was no glass on the ground at all.[11] The malleability of the reconsolidation process is such that it has even been used therapeutically to try to solve certain phobias or insurmountable fears.

False memories can also be created almost instantaneously, directly interfering with the consolidation process. This has been shown with the Deese-Roediger-McDermott (DRM) paradigm, an experimental procedure developed for a study of false memories that was based on the pioneering ideas of the scientist James Deese and published in 1959.[12] In the DRM paradigm, participants are verbally presented with a list of related words and asked to remember them, for example, "bed," "rest," "wake up," "tired," "blanket," "sleep," "snore," "nap," "peace," "yawn," and "sleepiness." After hearing the complete list, a few words are selected and the participants are asked if they remember them, but a decoy word is included among them. The decoy word is not on the list, though it does have some kind of semantic relationship with the other words. (In this case, the decoy word was "dream.") It probably will not surprise you to learn that approximately half of the participants thought that they had heard the decoy word.

Sometimes the information received after witnessing an event changes our memory of it, becoming a potential source of misinformation. This could be trivial if it happens to us with respect to events in our daily lives, but it could be fatal if it affects our testimony related to a crime. Elizabeth Loftus demonstrated that indeed the memories of witnesses to a crime can be contaminated when these people are exposed to media information containing erroneous details.[13] The realization that memories are so fragile has led to an initiative in the United States, the Innocence Project, which, thanks to DNA testing, is succeeding in exonerating innocent death-row inmates whose convictions were based solely on eyewitness testimony.[14] More controversial has been Loftus's defense of convicts accused by alleged victims who claim to have repressed memories of abuses they suffered during childhood. In reality, according to Loftus, these could very well be false memories implanted by suggestion in psychoanalysis or hypnosis sessions; the information suggested to these individuals may have been integrated into their memory as if it were an experience of their own.

As Loftus vehemently reiterates in her presentations, even a memory recounted in great detail, with confidence and emotion, can be false. As she herself concludes in her lectures, "Memory—like freedom—is something very fragile."

Memories and Memory Manipulation in Magic

Magicians want the outcomes of tricks to have a great impact on the audience—they want the outcomes to be the most memorable moments of the show. Magicians who are experts in the theory of magic, starting with Robert-Houdin in the nineteenth century, have proposed various ways to sequence the tricks in a show according to their impact. From what we know about memory in general, however, we must assume that the audience will remember very few details of the show they have attended. We have recently shown that memories of a magic show fade like any other memory. Over time, salient moments are remembered more than the tricks themselves, no matter the order of their presentation.[15]

A survey of more than five hundred people shared by the American magician Joshua Jay revealed that card games, although frequently mentioned, are in fact the least memorable kind of trick, probably because so much information is offered in card tricks that the public can follow the trick only if they pay close attention. Moreover, card tricks create situations in which memory is easily manipulated. In general, we remember what is easy to describe and understand. Spectators remember "Grand Illusions" best because it is stage magic with large objects. It is also true that tricks that combine cards with other elements are remembered better than tricks with cards only.[16]

Magic tricks are constructed using techniques for manipulating memories, either in the service of the trick itself or to prevent the audience from reconstructing the method afterward. Magicians try to interfere in the phases of acquisition, storage, and evocation of memories. Distractions can lead to forgetfulness and thus interfere with acquisition and consolidation. Then, in the evocation, or retrieval, stage, the reconstruction of memories can be manipulated by introducing disinformation and even by implanting false memories.

Scientific studies on disinformation and false memories—outside of magic—describe different types of manipulation procedures, most of which require quite a bit of time to create a false memory. In magic, these techniques are executed with similar results but in a very short time—the few minutes it usually takes to present a magic trick. The effectiveness of some disinformation maneuvers in magic parallels the DRM paradigm, through which false memories can be achieved almost instantaneously. As Juan Tamariz once said, alluding to his fellow magicians, we are experts in provoking gaps in memory, making audiences forget what we want them to and remember things that in reality did not happen.[17]

Techniques for the Promotion of Forgetfulness in Magic

Three techniques for promoting an audience's forgetfulness stand out, all of them based on diverting or dividing attention in order to alter the encoding and consolidation of memory:

Diverting attention. When attention is temporarily diverted during a trick so that the spectators do not consciously "fixate" on a certain maneuver, the encoding-consolidating process is weakened as a memory travels from short-term to long-term storage.[18]

The "parenthesis of oblivion." The parenthesis of oblivion, brilliantly described by Ascanio, is one of the maneuvers most used and celebrated by magicians to prevent the audience from "rewinding" to a particular moment in the presentation of the magic act.* The technique establishes a temporal distance between the critical moment of the trick and its related effect. For example, a magician hides a coin in plain sight, lingers for a moment—for example, by telling a short anecdote—then makes the coin magically appear suddenly in the same place. Through temporal dissociation, the spectator is prevented from establishing a connection between the tricky maneuver and the effect.

Recent experiments have investigated the relationship between the power of the magic trick and the time that elapses from the execution

* The parenthesis of oblivion corresponds to the "time and place dissociation" described by Tommy Wonder and is part of the English concept of "time misdirection" (Fraps, "Time and Magic").

of the trick to the outcome. If a little time passes, the trick can be easily discovered, but if a long time passes between execution and effect, the magic disappears because the audience tends to forget what was happening.[19] Magicians therefore have a limited window of opportunity—the parenthesis of oblivion—to influence the perceptions of their audience. These results were anticipated by Ascanio, who affirmed that the parenthesis of oblivion is effective only if not much time elapses between the initial situation and the final one; otherwise, an "anti-contrast parenthesis" is produced, causing the spectator to lose the thread of the exposition.

Sudden erasures of short-term memory. The promotion of forgetfulness can also be achieved by introducing highly distracting comic gags; these interventions can also be useful to resynchronize the audience's attention. A radical technique for doing this is Tamariz's "abrupt question," a good example of using distraction to avoid short-term memory storage of the moments prior to the action. To ask an "abrupt question," the magician quickly changes tone, in the literal sense of the word, and jumps in suddenly and energetically with a statement or question to break the discursive line. With their attention broken, spectators do not commit to memory an ephemeral event that occurred at the same time the magician made their outburst.

Let's take an example: Tamariz asks a spectator to name a card, and suddenly he turns to another spectator, loudly asking that person to put some energy into shuffling the cards. Obviously, the first spectator has already named a card, but Tamariz, having turned immediately to the other spectator, has not officially acknowledged this information. His sometimes shouted request that the second participant shuffle cards energetically breaks the previous discursive line, and thus the audience forgets that the magician asked what the card was.

Disinformation and False Solutions in Magic

In magic, disinformation can be used to induce forgetting quickly and very efficiently. Juan Tamariz is known for his ability to misinform the public by manipulating the recapitulation of events that happened during a trick, before the denouement or climax. For instance, he often, in a convincing albeit false way, insists to the spectator helping him with a card trick, "Isn't it true that you've shuffled and cut several times?" Another example of this tactic is provided by the magician Dani DaOrtiz in his trick called "The One."[20] DaOrtiz's constant patter manages to convince a spectator that she has freely thought of a card that he previously, and blatantly, forced on her. It would be interesting to know what the spectator really thinks about what happened at the end of the trick, and also what the audience really perceives. Again, these are matters well deserving further investigation.

In their small world of experts in deception, magicians have already lost their innocence and cannot believe that the audience is cognitively incapable of discovering their tricks. For some, introducing maneuvers so that the audience will not "rewind"—will not think about reconstructing the effect—is therefore a real obsession. They do that using different strategies. First, they execute the tricks in such a way that nothing provokes contrast with the logic of the exhibition—nothing attracts attention or induces the spectator to lose the thread of the presentation. As Miguel Ángel Gea states, "The aim of the magician is to deceive the unconscious and captivate the conscious."[21]

Second, they constantly lie to their audience. Early on, Robert-Houdin advocated in his "general principles of prestidigitation" for the need to "mislead" the audience to prevent them from mentally going back and discovering the mechanisms behind a trick.[22] The use of false clues is the great concept that Juan Tamariz develops in depth in *The Magic Way*.[23] Tamariz suggests creating false expectations or subtly suggesting solutions to the spectator throughout the trick that will later be seen as erroneous. Throughout the course of the trick, however, the magician succeeds in moving the spectator away from the real answer.[24] Spectators know that magic has secrets, and according to the Dutch engineer and magician Peter Prevos, when faced with the cognitive dissonance that comes at the climax of a trick, they are "biologically" impelled to think about the method so as to recover the cognitive control that has been disrupted by the magician.[25] False clues create additional distance between the spectator and the "real" answer.

In practice, false clues sometimes consist only of brief maneuvers, such as making a simple gesture at a precise moment while handling a deck of cards. At other times, the performance is deliberately chaotic, seemingly careless and sloppy, like the shows of the Swedish magician Lennard Green. Others, like Dani DaOrtiz, falsely pretend to improvise at every step, in order to relax the audience. All in all, even subtle, implicit actions are often enough to mislead and create the artful effects of a good magic trick.

Third, false solutions may also prevent the public from prematurely arriving at disappointing conclusions about the method behind the trick. For the American magician Darwin Ortiz, this sudden insight is known as the "aha! effect"—a moment that ruins the magic trick because the layperson believes that they have discovered how the method works, even though their deduction may be erroneous.[26] It has been observed that aha! moments occur more frequently than we think. At such moments, spectators have high confidence in the solution they have found, for when the mind already regards an idea-solution as reasonable, it is difficult for alternatives to be considered. This aspect of reasoning is called the "Einstellung effect," which was described experimentally in 1942 and has been replicated on numerous occasions ever since.[27] These moments are therefore happy ones, and as a result, the solution they imagine—especially if it is correct—is memorable even weeks later.[28]

If the illusion of the impossible is not compelling enough, spectators will be predisposed to try to understand maneuvers or to look for solutions, consciously thinking about what they have seen. At the other extreme, as Juan Tamariz states, when one achieves "the spell and the fascination" that generates the illusion of impossibility, one can end up "suspending both the desire and the capacity to search for the truth about what one has experienced, since one would not even know where to start."[29]

Some spectators may notice the moment the magician executes the method, but in fact that rarely helps them figure out how the trick is engineered.[30] Moreover, it has been observed that laypeople tend to overestimate the ability of other spectators to deduce the method behind the trick, especially if they have discovered it themselves.[31] In short, magicians may to a certain extent overestimate the audience's ability to detect their methods.

Long-Term Memories of a Magic Show

How might a magician manipulate long-term memories to distort what the spectators take home with them? It is not uncommon for someone to imagine that they witnessed a magic trick decades ago and to recount it in great detail as a great experience. It is also true that what spectators relay about their experiences often differs completely from what really happened, because, as we have discussed, the brain retains only conceptual simplifications and the evocation of memories is a process of reconstruction. What is remembered and narrated changes over time; the reconsolidation of memories can be influenced in very subtle ways, resulting in the addition of elements that never happened.

Juan Tamariz uses the term "comet effect" to refer to this "tail," or memory, of the magic that lingers with spectators long after a show. He maintains that "a good trick should be a comet . . . followed by a long tail," and that, to achieve this comet effect, the magician must work during and after the show with the objective of enriching the memory, manipulating it if needed. The magician needs to create a "fantasy narrated by the spectator," because the spectator will need to "revive or transmit the sensation of miracle, of wonder," that he has experienced, exaggerating "the final effect and the conditions." In his last work, *The Magic Rainbow,* perhaps his most extensive treatment of the comet effect, Tamariz proposes diverse "boosters of positive memories," such as the "symbol," the "persona," or the "evoking hooks." These techniques aid in the creation of inaccurate or totally false memories, among which he includes "impossible promises" or facts that later are not fulfilled—such as, "I won't touch the cards again"—as well as techniques for the erasure of memories that we have already discussed.[32]

The Australian Richard Hodgson, a researcher of paranormal phenomena, recognized back in 1887 that the memories of the spectators of a psychic seance were not very reliable.[33] Later on, in 1932, Theodore Besterman, a Polish bibliographer specializing in psychic research, conducted a controlled experiment in which he reproduced a performance of paranormal phenomena before forty-five volunteer witnesses.[34] Asking these attendees for details after the session, Besterman was surprised to find that the accuracy of their testimonies ranged widely, from 5.9 percent to 61 percent, but on average was only 33.9 percent, and that the memories retained were otherwise vague and imperfect.

Besterman was a contemporary of Harry Houdini (1874–1926), the famous American magician of Hungarian origin who became very famous for his escape spectacles, and of the even more famous Scottish writer Sir Arthur Conan Doyle, creator of the character Sherlock Holmes.* Besterman, known for his skepticism and for discrediting believers in psychic phenomena, earned himself a great enemy in Conan Doyle, who professed great faith in this type of phenomena—to the point of believing that Houdini's escapes were real and that the magician could in fact dematerialize, even though Houdini personally explained his tricks to him.[35]

Contemporary studies have determined that people who believe in paranormal phenomena seem to have a worse memory for details that are re-created in experimental situations and, in turn, are more likely to remember events that never happened.[36] They are also much more susceptible to manipulation through suggestion and instruction, especially when the former is congruent with their belief in the paranormal.[37] This is the well-known "confirmation bias," a cognitive bias that describes our tendency to favor information that confirms our own beliefs or hypotheses.[38]

The memory that a spectator retains of a magic show is likewise extremely variable and depends on personal experience, the spectator's passion for magic, the tricks executed, and even the magician's appeal.

In the survey mentioned earlier, Jay asked 526 people (mostly Americans, 28 percent of whom had never seen a live magician) what they liked best about magic. Most participants said that it was the desire to

* Houdini's real name was Erik Weisz; he took on his stage name to honor his predecessor, the French magician Jean Eugène Robert-Houdin, the father of modern magic.

be amazed and to enjoy the surprise of magic. On the other hand, only 10 percent answered that it was to "know how the tricks work."

When asked what they liked least about magic, 14 percent reported that they disliked not knowing how the trick was done, while 16 percent reported that they felt annoyed if they found out how the trick was done. When respondents were given a choice between watching magic or seeing how magic works, 60 percent preferred the former.[39]

Audiences' responses are diverse and very personal because, although it may seem otherwise, a magic show tries not to engage our ability to reason. On the contrary, as we will see in the next chapter, audiences remain unconscious of magicians' maneuvers most of the time.

CHAPTER 9

The Undervalued Unconscious Brain

The Brain Never Rests

It's early morning and you are off to work. As you go down the stairs and out into the street, you take the first bite of a sandwich you hold in one hand and leaf through the headlines of a newspaper you hold in the other. You continue walking, and with your mind fixed on the tasks your boss assigned you yesterday, you arrive "without thinking" at the bus stop.

To achieve what we have just described, the brain must perform thousands of operations per second. But we are not conscious of this happening, as these operations are done in a learned, automatic, and unconscious way. Generally, we probably overestimate the role of the conscious brain; most of what we feel, think, and do is not under its control. It is the unconscious brain that allows us to survive in a world that requires us to constantly process a huge amount of information.*

It may seem that going through our day perceiving through our senses, invoking memories, reasoning, and making decisions requires no effort, but in reality the opposite is true: all of these operations require huge and constant effort. The brain is constantly filtering, processing, integrating, and storing information, but most of this activity is unconscious—including what happens while we are sleeping. In fact,

* Although for practical reasons in this book we distinguish between the conscious brain and the unconscious brain, the reality is that there are physical parts of the brain that can process information in both ways. We are therefore making a distinction that is more functional than anatomical or structural.

the unconscious machinery of the brain is responsible for most cognitive functions, in addition to taking care of vital functions such as controlling breathing, heart rate, temperature, and insulin secretion.

The unconscious brain solves the great puzzle of reconstructing reality and predicting the immediate future. In an uncountable number of calculations per second, it is constantly making automatic and instinctive decisions. However, we are aware of only the results of the brain's operations, not of the operations themselves. To paraphrase the neuroscientist David Eagleman, it is as if "the brain runs its show incognito."[1]

Attention and Awareness

When it comes to the relationship between attention and awareness, there are two primary schools of thought. One holds that only objects attended to become conscious, and that only objects consciously perceived can be attended to; in other words, attention and consciousness are intimately related.[2] The other school of thought maintains that attention and consciousness are interrelated but separable processes with differentiated functions and distinct neural mechanisms.[3]

Experimental data suggest that the relationship between attention and conscious perception is complex. Most attention mechanisms are unconscious, and not all things that we attend to reach consciousness. For example, an individual who tries to concentrate on the sounds coming through the earphone of one ear while trying to ignore the sounds coming through the other earphone is nonetheless unconsciously aware of many of the sounds that reach the unattended ear. These sounds leave certain traces in the brain that can influence what is being paid attention to through the other ear.

In addition, there are stimuli that become conscious even if they have not been expressly attended to—as happens, for example, when stimuli (such as faces) are identified very much on the periphery of the visual field, almost in the absence of attention.

Finally, attention can be directed to something that has not been consciously perceived but that may have been previously conditioned. This is what happens after "priming," as long as we have first immersed ourselves in the world of implicit memories.

Attention without Consciousness

You have already left the house; you are eating a sandwich, leafing through the newspaper, and heading toward the bus that will take you to your workplace. As you walk to the bus stop, you notice neither all the houses nor all the people passing each other along the way, nor all the cars driving down the street. You approach the bus stop with your mind intent on different topics, and you will only be struck by what contrasts because it is either unexpected or notably different from your daily routine: a noise, a smell, or a person dressed in a striking way, for example. You arrive at the office, and after a while you welcome someone you have been scheduled to meet but have never met before. The person looks familiar, but you do not know why. During your conversation, you discover that he lives close to you and that he has taken the same bus that you have. You had not "noticed" this person, nor was there any special reason to do so, but in your unconscious, you had registered his existence.

Whether or not a given stimulus rises to the level of consciousness depends on two primary factors: the attention devoted to it and the intensity of the stimulus. But there are also stimuli that are perceived subliminally—in a not totally conscious way—either because the stimulus is weak or attention is directed elsewhere. Such stimuli can provoke the creation of "subliminal images": those images or words perceived below the threshold of consciousness.

Experimentally, it has been shown that subliminal images can be induced. The scientist Stanislas Dehaene has carried out experiments using the "masking" technique, which consists of very briefly presenting an image or a word preceded and followed by similar images or words (for example, a series of random geometric shapes) that act as masks and end up making the original image "invisible" to the spectator's consciousness. This happens because the difference between "seeing" and "not seeing" an image or word lies in the duration of its presentation. Numerous experiments have shown that, if the image or word is presented for 30 milliseconds at most, it is "invisible," but if the duration is increased to 60 milliseconds, then the word or image will be easily seen or heard.

The threshold of "invisibility" varies between individuals, but even so, it is always between 30 and 50 milliseconds. Interestingly, as Dehaene's work has suggested, if the image is a smiling or frightened face, then participants subliminally (unconsciously) perceive a similar emotion. This happens because visual information takes a quick shortcut to the amygdala, the nucleus of the brain involved in, among many other things, the conscious and unconscious processing of emotions.

This is how masked images can be unconsciously recognized, deciphered, and interpreted. It is important to emphasize, however, that subliminal perception is ephemeral compared with conscious information. In conscious processes, and thanks to short-term memory, a visible image can be remembered for a certain period. Subliminal thoughts last for only brief moments, and although it is true that they influence our thoughts, this influence is also ephemeral.

Dehaene's experiments have also shown that masked images can still activate visual, orthographic, lexical, or even semantic representations.[4] Though this unconscious activation will quickly fade, is it possible that there is enough time in certain magic tricks for spectators to unknowingly capture some subtleties of the methods performed by the magician?

Unconscious Perception in Magic

We now know that information not consciously attended to can be processed unconsciously. But how do magicians use this information?

For Miguel Ángel Gea, this fact is as true as it is relevant because, according to him, the hidden techniques used by magicians are sometimes not so imperceptible and may be "felt." That is, though they are not likely to be consciously felt, yet "they reach the spectator's brain in such a way that they end up affecting the experience of the trick" to the extent that, as Gea recognizes, sometimes the spectator "senses the trickery, even though he or she doesn't perceive it."[5] This spectator is receiving information in an unconscious way that conditions the "experience of the impossible."

If this is true, then a new question must be asked: Are there unconscious forms of attention and perception that depend on how the magician constructs and presents a trick?

The response to this question would appear to be yes. Researchers recently observed that, during a magic act, members of an audience were able unconsciously to detect the magician's maneuvers, or subtleties supposedly hidden or invisible.[6]

Diego Shalom and his colleagues have described physiological changes—transitory dilations of the pupil—related to spectators' subliminal perception of a card that they have seen for a little longer than the others when a magician riffles the deck in front of them.[7] This tricky riffling of the deck goes unnoticed by spectators, but it influences their decisions, thus helping the magician force the desired card. Often in magic tricks, so much information is conveyed that it cannot all be consciously attended to, but bits of information leave an unconscious trace in some spectators, which can influence their perception of the trick.

Moreover, unconscious perception can explain why the same effect performed by different magicians does not necessarily achieve the same magic experience. In fact, even when the effect is supposedly done in the same way, with the same techniques and the same procedures, subtle differences may lead to different experiences from the trick—different enough at least so that, in one case, "the experience of the impossible is a great one," as Gea would say, and in others it is not so magical. Something similar happens when, for example, laypeople affirm that they "notice" something in the performance of singers or instrumentalists when they are off-key, but they are not able to say exactly what that is. In magic, if the magician is "on-key," the magic experience will be magnificent, but if the magician is off-key, the experience may not be as amazing. And a different magician, even using the same trick and the same techniques, may cause the spectator, unaware of any differences in how the trick was performed, to capture different information subliminally, at an unconscious level.

Implicit Memories

Besides the explicit and conscious memories that we discussed at length in the previous chapter, there are the other important long-term memories called implicit memories, which are unconscious. Implicit memories are those memories that retain learning, habits, and conditioning; unlike explicit memories, they are more reliable, fixed, and lasting.

We are born with some of our implicit memories. The phylogenetic memories, typical of our species, are coded into the genes of each individual and guarantee the basic functions necessary for our survival, such as the sucking reflex in babies or our reactions to threats.

Implicit memories can be subdivided into motor, or procedural, memories and perceptual, or emotional, memories. Motor memories are formed in the striatum outer section, in the basal ganglia, and in the cerebellum, the great repository for all types of learned procedures. When a behavior or motor process has been learned, it is stored in these subcortical structures in the form of an automated program that allows the brain to perform complex tasks with minimal expenditure of time and energy.

Your motor memories include learning how to dress, tie your shoes, ride a bike, drive, play an instrument, or type on a computer keyboard. If these kinds of skills are not well honed, they cannot be carried out automatically, so you cannot perform them while having your mind on other things. When something is first being learned, it forms part of your explicit memories, which are of a conscious nature; then, with repetition, these learnings are consolidated as implicit memories and become unconscious and automatic. Thus, when you are learning to drive or ride a bicycle, you must be totally concentrated on that activity. Once you have learned and mastered the skill, however, you can ride a bicycle while admiring the landscape or drive a car while talking to a passenger. This type of implicit memory is popularly known as "muscle memory."

In magic, thanks to many hours of rehearsal, some magicians manage to perform several tasks simultaneously, an achievement that can be a show in itself. For example, sometimes a magician can dance while juggling, or perform sleights of hand with a deck of cards while asking the audience questions so that they do not notice the maneuvers.

Curiously enough, unlike voluntary movements or behaviors, which are under the control of circuits in the cerebral cortex, automated procedures, once learned, are guided at a subcortical level, without conscious control. In fact, if you stop to think about an automated procedure in a conscious way, you become discombobulated and cannot accomplish the task correctly. For unconscious automatism to function properly, it is better to be distracted, to avoid thinking about the steps or components of the learned task. For this reason, many athletes who are well trained to enter the "zone"—or "automatic mode"—need to carry out certain rituals that help them reach a less conscious state. Before a race, for example, some athletes will perform a series of routine movements that are not meant to "bring them luck" but rather to prepare them to perform in the best possible way, "without thinking about it."

The other type of implicit memory is perceptual, or emotional, memories, which are controlled by conditioning mechanisms (such as fear) that involve the amygdala in connection with different cortical areas. Emotional memories include various habits and conditionings, such as a fear of snakes (which we inherited from our ancestors), addiction to a drug, or an obsession with an odor or spice. Some emotional memories can develop either sensitivity or intolerance—two different, opposite phenomena. In sensitization, the reiteration of neutral or weak stimuli makes them harmful or unbearable over time, such as when you end up hating a popular but overplayed song. Tolerance or habituation results, on the other hand, when the repetition of a certain stimulus either causes its effects to fade or renders them harmless. Tolerance has happened when you listen to the same song so often that you eventually stop consciously hearing it and do not even realize when it is playing on the radio.

Emotional memories are built through associative conditioning, either classical Pavlovian conditioning or operant conditioning (punishment and reward). In either case, the emotional memories acquired through repetition take root in us in such a way that they also give rise to automatic and unconscious behaviors. A friend of ours who had the habit of smoking while talking on the phone confessed to us that he still has to suppress the urge to get a cigarette every time he picks up the phone at the office.

Most experiences combine both explicit and emotional memories in a simultaneous and complementary way. It is precisely the implicit, unconscious character of emotional memory that makes modifying it a very complex task that requires creating a new habit to replace the old one.

Subtle Conditioning: The Case of Priming

It is possible to influence attitudes, perceptions, and choices in an unconscious way. This phenomenon, called priming, is related to implicit memory: it works because previous exposure to certain stimuli (auditory, visual, tactile, and so on) can influence our response to other stimuli presented later. The effects of priming are natural and automatic, and our daily experiences are full of them. Priming maneuvers unconsciously activate specific associations before we carry out an action or task. For example, if we see the word "yellow," it is easier for us to later recognize or evoke the word "banana," a conceptual priming that occurs because "yellow" and "banana" are closely associated in our memories.

Priming is more effective when both stimuli share the same sensory mode. For example, visual priming works best on visual cues, and verbal priming works best on verbal cues, because not only are the target stimuli primed conceptually, but the neural circuits involved in detecting and processing the target are primed and preactivated.

We can distinguish perceptual priming from conceptual priming. Perceptual priming takes place between items that have a similar shape and is favored by the degree of similarity between the initial stimulus and the subsequent one. For example, if a person is presented with a list of words that includes the word "tablet" and then asked to participate in a word completion task, the chances are much greater that, when presented with the letters t-a-b, they will give "tablet" as a response than if they had not previously seen that word on the initial list. Let's see now what happens when we try to complete the word "so—p." If we have recently heard the word "wash" or seen things related to hygiene, we will tend to complete the word as "soap," but if we have been talking about restaurants or gastronomy, the word is easy to complete as "soup."

Conceptual, or semantic, priming occurs between items with a similar meaning and is revealed, as its name implies, during semantic-type tasks. For example, the word "table" will show priming effects on the word "chair" because both belong to the same semantic category of "furniture." The word "dog" may be priming for the word "wolf" because they are similar animals. Morphemes, even though they are not words, can produce a priming effect on whole words in which they appear.

Semantic priming takes many different forms. Subliminal semantic priming, for instance, is induced by masking techniques, as we have already seen: it is achieved by projecting for a short time a subliminal word or image followed immediately and very briefly (less than a second) by an item that is visible.[8] The previous presentation of this word or image unconsciously accelerates the subsequent processing when that same word reappears in a conscious way.

For example, people respond faster and make fewer mistakes when the word that precedes and is presented subliminally matches the word that is visible because, as the cognitive neuroscientist Dehaene explains, "much as one primes a pump by flushing water into it, we can prime the circuit for word processing by an unseen word."[9]

Priming in Magic

In some magic tricks, magicians will surprise the audience by "predicting" the choice that a spectator will make. Many experienced magicians are capable of achieving priming effects during the few seconds or minutes that the presentation lasts, influencing through direct or indirect clues (color, suit, value) the spectator's subsequent choice. The key to priming is to know that a spectator invited to make a decision or choice will generally choose according to the clue that they have been exposed to the longest.

The mentalist Derren Brown practices a subtle priming technique using finger signs. Making "mental pictures" of cards with his signs, he claims to be able to force, for example, a choice of the two or three of diamonds.[10] Brown is a magician who likes to give spectators "scientific" explanations, but this is part of his theatrical script. In 2014, on his popular television show *Mind Control*, he invited a spectator to mentally choose a toy from an imaginary store; once the spectator had chosen, the mentalist guessed that the selected toy was a giraffe. Brown claimed that he had used a priming technique to subtly direct the spectator's mind to a toy giraffe: he presented the symbol of a giraffe with his hands while giving instructions to the spectator, who, unaware of having been induced to make that choice, felt that he had made a free choice.

In everyday tasks, the unconscious mind is the primary source of human behavior. It also plays a big role in how we perceive and experience reality. It is not surprising, then, that magicians have learned to take advantage of the unconscious functioning of our brains to perform impossible tricks that amaze us and challenge our understanding. In the next chapter, we will discuss how, without our knowing it, magic hacks our unconscious brain to bias our behavior and condition our decisions.

CHAPTER 10

The Magic of Decision-Making

The Dual Functioning of the Brain

It has long been accepted that two large systems enable the brain to make decisions. One is the domain of the unconscious brain, which is automatic and intuitive, and the other is the domain of the conscious brain, which is analytical and supposedly more rational. Many models of dual information processing have been formulated, such as those by Freud and by Pavlov.[1]

The models of dual processing distinguish between the automatism in the unconscious functioning of the brain and the mechanisms and reasoning of which we are aware. Daniel Kahneman, the Nobel Prize winner in economics in 2002, proposed a distinction between system 1, the automatic system, or the unconscious, and system 2, the rational system, which is conscious.

System 1, the automatic system, quickly and efficiently directs most everyday behavior. System 1 is an unconscious system with an associative, effortless, slow-learning and automatic nature; being influenced by the subject's emotional state and previous experience, it is very sensitive to context as well as subjective. According to the neuroscientist Antonio Damasio, not enough importance has been given to feelings as modulators of decisions and other automatic reactions.[2]

Requiring few cognitive resources to function, system 1 often has a great operational capacity at any given time. Some examples of the processes carried out by system 1 would be adding 2 + 2, perceiving that an object is farther away than another, or driving a car on an empty road.[3]

System 1 is also resistant to change, learns slowly, and can be conditioned or primed. It is quick to react. but not necessarily irrationally; many immediate decisions involve processes that, before being learned, were once very complex. System 1 usually works perfectly, although as we will see, it can make mistakes in situations of uncertainty.

System 2, the rational and conscious system, is a highly controllable processing system, but it is also lazy and slow. System 2 is flexible and malleable through reasoning and is linked to language. It demands large amounts of cognitive resources and is limited by the capacity of working memory, which is short-term. Processes executed by system 2 include, for example, identifying an unusual noise, walking at a faster pace than is natural, parking in a narrow space, or comparing two appliances to determine which is better.[4]

In decision-making, and in contrast to system 1, system 2 is responsible for questioning and even correcting—but not suppressing—the operation of system 1. System 2 is therefore an analytical system that requires a conscious effort. And although it can be very reflective, sometimes great mistakes are made in decisions that have been pondered for a long time.

When we have to opt for one of the two systems to solve a problem, which one do we choose? Kahneman has empirically proven that most people feel comfortable accepting the immediately available solution provided by system 1. But the reality is that we probably do not select one system or the other, just as we do not select conscious or unconscious mental processes. There are no clear boundaries between the two.

For instance, we arrive at many stereotypes and biases automatically, although they are not as unconscious as we would like to believe. As happens with any simple and intuitive model, the dual-processing model, "system 1 versus system 2," has been used in a very reductionist way and has therefore been the object of criticism.[5] Kahneman himself, in fact, insists that his proposal is not a theory but a metaphor that helps to describe and understand how we think.[6] In reality, systems 1 and 2 influence each other as they work in tandem, either sequentially or simultaneously. But we do function most of the time rather automatically, which is our default.

From this perspective, magic always seeks an ally in system 1. When magicians ask questions to prompt choices, they systematically look for automatic answers, which are not reflected on and therefore do not activate system 2. Along these same lines, the magician's objective when presenting a trick is to have spectators proceed without blinking an eye and without stopping to think about how the maneuvers are taking place. What the magician seeks, in other words, is for spectators to stay in system 1 and not switch to reasoning with system 2. If this happens, spectators will start to ponder on what they have observed and then probably lose the thread of the presentation and miss out on the surprising final effect.

Do We Make Expert Decisions?

Most of our daily behavior is unconsciously controlled. Without realizing it, we live in the illusion that we rationally "control" many more things than we actually do—for example, when driving a car in very heavy traffic. Within the framework of system 1, most of our decisions are instinctive, quick, and intuitive judgments made, with little or no conscious effort, by relying on our experiences and memories. Nonetheless, our feeling of being in control is the same whether we make decisions consciously or unconsciously, since that feeling is a postdictive construct—an illusion analogous to the perceptual or cognitive illusions discussed earlier.

The simplicity of unconscious decision-making presents advantages, especially in situations of uncertainty. Under these conditions, heuristics, or general rules capable of predicting complex phenomena, can work as correctly as, or better than, more elaborate rules. For instance, when a lot of information is available, it is more difficult to compare alternatives, and the possibility of conflict increases. This was proven in an experiment as simple as observing that customers will buy more jam if they are given the opportunity to choose between six varieties instead of twenty-four.[7]

At first glance, consistently relying on unconscious processes to make decisions would seem to be irrational behavior, but is it always better to make reasoned decisions? To what extent can we pay attention to "first impressions"? The answer is that it depends, because, when faced with decisions on matters in which one is an expert, system 1 is apparently at least as effective as system 2.

Recently, Keiji Tanaka has confirmed this, based on the choices made by players of shogi, a two-player Japanese strategy board game akin to chess; unlike in chess, however, it is easier to tell in a game of shogi whether a player is following an offensive strategy or a defensive one. Tanaka did his research with two types of players, experts and novices, and in two different visualization conditions: a long presentation, lasting seconds, that gave players time to ponder the answer, and a short presentation, lasting about half a second, that only allowed players to answer intuitively. Both experts and novices were presented with a game position and asked to propose their next move. Tanaka observed that, when given enough time, both expert and novice players made their decisions using the same regions of the cerebral cortex, which were as optimal as they could be in relation to their level of experience in the game. However, when the position of the game was presented subliminally, lasting only half a second, the novice players responded almost randomly, using the same cortical areas as in the first scenario, while the expert players made very good decisions using different brain structures, such as the basal ganglia and the cerebellum, but without being aware of how they had done it.[8]

In response to the question of whether we can trust our intuition to make good decisions, we can say that it depends on our level of expertise in the particular area.

Judgments in Situations of Uncertainty and Instinctive Decisions

Instinctive decisions are made by heuristics, those shortcuts or learned rules that allow us to solve problems, make decisions, and act quickly with very little information. Heuristics have had a great evolutionary value because surely, without such shortcuts, the first hominids would have been devoured by all kinds of predators.

Some heuristics are automatic, and others are more flexible. Recognition heuristics and gaze heuristics, for instance, support rapid evaluation and what has been called the "intelligence of the unconscious."[9] On the other hand, heuristics can sometimes lead to imperfect evaluations and decisions, known as cognitive biases.

Daniel Kahneman and Amos Tversky have described some of the consequences—sometimes even errors—caused by the unconscious and automatic use of heuristics when situations of uncertainty lead us to make instinctive judgments.[10] When the context or circumstances change, our instinctive decisions may no longer be adaptive. In addition, Kahneman and Tversky have demonstrated that human beings make automatic decisions without taking into account the laws of probability—for example, when we invest large amounts of money in the lottery when the probability of winning is much less than that of dying, or when we cannot stand hearing a music playback system repeating the same piece we have just heard. In the latter case, Apple had to modify the iTunes random-listening function because customers refused to tolerate hearing the same song repeated—which, logically enough, can happen if the selection is truly random.

A magician takes advantage of our inability to correctly understand the concept of chance. After the magician shuffles a deck of cards so that the card selected by a spectator will get "lost" in it, they always make a point of saying that since the card is really lost, it is neither on top of the deck nor at the bottom, and they openly show those cards faceup. However, if the magician had really shuffled the deck randomly, the "lost" card could indeed have turned up by chance on top or at the bottom.

In fact, human beings do not seem to instinctively understand the full meaning of randomness, even though chance in nature is much more frequent than we usually recognize. We have evolved to detect patterns that allow us to make sense of things and thus maintain "our" order.

One of the consequences of our training in detecting patterns is the ease with which we establish cause-and-effect relationships, often of an illusory nature; many people are thus easily captivated by conspiracy theories. Similarly, as we easily find meaning and "rationalize" instinctive decisions, we avoid having to attribute any fact or act to chance.

The truth is that, in situations of uncertainty, we commonly resort to general heuristics, as Kahneman and Tversky have described in detail. Take the anchoring heuristic, which explains our sometimes absurd behavior when we discuss a price in a bargaining match. Why do we begin with the seller's starting price and not with the supposed real value of what we want to acquire? Another would be the representativeness heuristic—our tendency to judge the probability or frequency of an event or fact according to how well it fits a prevalent stereotype or received knowledge. The representativeness heuristic is at work when we make biased judgments of other people simply because they belong to a certain ethnic group or community. We also have the availability heuristic, which is a type of shortcut that magicians take advantage of when they induce choices. Let's see how it works.

With the availability heuristic, we estimate the frequency or importance of a category in relation to the images, concepts, or ideas that first come to mind, or in relation to the information available when we make that assessment.[11] A classic example would be deciding whether to take a plane the day after a serious air accident. In reality, this decision, although it may be unbearable for a person who tends to be anxious, should not take into account the last accident, because the probability of another one occurring is minimal—that is, it remains the same as always.

The availability heuristic is what underpins the "prototypicalities" in all kinds of categories—that is, how representative a subject or an object is of a given category, such as animals, colors, and the like. The prototypical examples of each category, the ones we think of first, will vary for each of us according to age and gender, but primarily according to our culture.

These are the main rules or shortcuts that allow instinctive decisions to be made quickly and with little information. As we will see later, it is precisely these instinctive decisions that magicians prefer when they are soliciting answers or decisions from spectators and when their tricks expressly provide little time and little information. This maneuver is called "forcing"—pressuring the audience for a question or choice and giving them no room for reflection in order to ensure predictable answers or decisions or to exercise total control over the audience's reaction.

Forcing does not necessarily always involve inducing the spectators to make a choice.[12] Sometimes the magician manages to ensure that the option he wants is the chosen one by simply controlling the choice set—that is, the possibilities available to the spectator at any given time. In any case, it is of paramount importance that the magician always make sure that the spectators feel that they have chosen freely.

One of the secrets of forcing is that it always involves a certain amount of pressure. The spectator is induced to make a decision or to answer a question with very little time or with very little and often distorted information—in other words, without the opportunity for introspective deliberation.[13]

As we saw in the previous chapter, magicians are capable of prompting a certain choice in such a way that spectators are unaware of being manipulated and convinced that they have made a completely free choice (thanks to techniques that, by the way, are also used by many salespeople). Scientific studies on magic tricks have found that some forcing techniques render spectators unable to distinguish between a true forcing and a free choice.[14] Studies have also shown that, under these conditions, it is not difficult to influence the sense of freedom with which decisions are made.[15]

Types of Forcing

In magic, there is a wide spectrum of forcing tactics aimed at manipulating spectators' decisions or responses, ranging from automatic and safe procedures to risky ones. If the procedure is well executed by the magician, some forcing techniques always lead to the desired result. There are other techniques that aim to get probable answers and whose result is not assured beforehand. These fall under the umbrella of what magicians call "psychological forcing."

In card magic, for example, there are many automatic techniques; those based on mathematical procedures stand out. Take the following example:

> In a deck presented from the back, you want to force a card that has been placed in the tenth position, counting from the top. To do this, the spectator is invited to say a number between ten and twenty. He is then asked to discard, one by one, that many cards from the top of the deck. The new pile of cards produced is placed on top of the deck and squared up. Then the spectator is asked to add up the two digits of the number he chose and to discard again, one by one, as many cards as the result of that sum. The final card after this new discard will be the forced card.

More complex techniques require skill and training; they depend on the skill of the magician, or they may require special elements. Among them is the "Magicians' Choice." Using this forcing technique, magicians control and vary the type of questions they ask while the spectator is making a choice. If executed well, magicians always manage to obtain the desired choice. Classically, this trick consists of presenting a scene with several options, such as several envelopes, cards, objects, numbers, words, or colors. The magician leads the spectator to give the desired answer or to make the right choice by asking ambiguous questions. The magician plays with the advantage of having several "outs" in the form of alternatives or new "logical" questions, depending on how the choice process unfolds. Let's look at the following example:

> The magician puts two cards on the table, facedown. One of the cards (the one to the right of the spectator) is the one the magician wants to force, that is, the one they want the spectator to choose. In general, right-handed people tend to choose the card on the right. To avoid mistakes or hastiness, at the outset the magician tells the spectator to point at or indicate a card, but the magician will never direct the spectator to pick up one of the two cards. If the spectator points to the card on the right, the magician tells them to pick it up. Goal accomplished. If the spectator points to the card on the left, the magi-

cian will say that it is the discard and will remove it, leaving only the card that is to be forced on the table. The important thing is that the spectator is convinced that their choice has been completely free.

Other forcing techniques capitalize on rules of perception. In chapter 6, we discussed Alfred Yarbus's demonstrations of our tendency to focus on the salient aspects of an image—the contrasting ones—and we learned, too, that the information processed after attentional fixation depends on the tasks or instructions we have received. Take the following example:

> A spectator is invited to think about a card while nearby the magician riffles a deck of cards. The magician exposes the cards one by one, faceup, and quickly (in fractions of a second)—except for the one they want the spectator to choose, which is the card exposed for the longest time (and which will generally appear between the middle and the end of the procedure).

This technique is successful close to 100 percent of the time, and most participants do not realize that the card has been forced.[16]

"Classic forcing" is the most widespread type of psychological forcing in card magic; it has been practiced with little variation since at least the nineteenth century.[17] Classic forcing is completely dependent on the training and experience of the magician, and thus it can fail. Let's see how it works.

The magician invites the spectator to choose a card from a facedown deck that has been laid out in a fan. The choice appears to be free, although in reality, as the magician surprises the spectator by pressuring them to make the choice quickly, it is a transfer that goes unnoticed. Once the spectator has picked up the allegedly chosen card, the pressure abates and the atmosphere changes; the magician emphasizes (explicitly or implicitly) that the choice has been free and usually shortly thereafter reinforces this statement again so that later the spectator will remember it that way.

Magicians who are skilled in classic forcing know how to create an atmosphere of trust and provoke spontaneous reactions by giving the spectators little room to think. As one of its greatest exponents, Dani DaOrtiz, states, "The spectator is not your enemy, let him win."[18] Though no one knows precisely why this technique works, we do know that the magician's attitude and cadence, the brief narrative of the instructions, the nonverbal communication, and, crucially, the level of the magician's experience are all very important. The seemingly natural behavior of the magician encourages the trust of the spectator, relaxing their attention and reducing their suspicion. All that remains is for the magician to choreograph the spectator's movements so well that the spectator does not realize that they have chosen precisely the card that the magician has wanted to transfer to them.

As the Spanish magician Tino Call notes, a classic forcing performed with the deck of cards fanned out faceup creates an even less suspicious situation, as this presentation reinforces the idea that the card is freely chosen, even though it is still induced.[19]

Controlling the spectator's reaction time is always central to classic forcing. DaOrtiz "manages the reaction time" of the spectator verbally, using a variant of classic forcing called "stop forcing." When he begins to discard the cards facedown on the table, DaOrtiz invites the spectator to say, "Stop!" and then counts out three more cards. The spectator usually tells him to stop at the fourth card.[20] Nonverbal communication is also decisive in this type of forcing. DaOrtiz highlights the importance of forcing maneuvers such as the gaze, the movement of arm and fingers, and, more concretely, approving nods.[21]

Taking a Risk

In riskier techniques of psychological forcing, the magician tries to prompt the spectator into a very specific type of reaction or answer. Because this type of forcing can fail, the magician always has a plan B—several potential outs should the spectator choose the "wrong" option. The first strategy of this type of forcing is to rely on previous priming to induce a choice and thus anticipate a particular answer—for example, to rely on a certain suit being chosen from the deck after priming it in some way.

A second strategy is to go after certain very probable responses that occur automatically, such as, in our example of "magician's choice," spectators' tendency to choose the stacks of cards or objects on their right. In a now-classic experiment, Richard Nisbett and Timothy Wilson invited customers visiting a department store to choose a pair of nylon stockings from a display in which several pairs were exposed in four drawers: although the pairs of stockings were of the same quality, the pairs that customers chose most often were those placed in the drawers to their right.* In a later interview, however, none of the customers admitted that their choice had anything to do with the stockings being on the right side of the display and instead offered irrelevant arguments, such as that the stockings they chose were softer.[22]

Tricks in card magic are based on similar procedures. For example, the magician places four or five cards or stacks on the table and invites (at his or her own risk) the spectator to choose a previously desired card or stack. Experience shows that, in this context, the spectator usually chooses the card or stack located in the second position, counting from the right if the spectator is right-handed. (Left-handed spectators will tend to choose the opposite.) Sometimes, to favor the intended choice, the magician will place the stacks or cards somewhat diagonally.

Why the second stack on the right and not the first? It is usually argued that the spectator uses an accessibility heuristic to choose and takes the stack closest to his right hand; if so, however, he should have chosen the first one and not the second one, especially when the stacks are arranged diagonally. What could be happening is that the spectator uses the accessibility heuristic, just as Nisbett and Wilson described, but being aware that the magician is trying to trick them, the spectator decides not to choose that first stack on the right and jumps to the second one. In any case, as the result may not be the expected one, the magician must have a plan B, perhaps by following the trick with proposals more typical of the "magician's choice," which always leads to the desired result.

*Forty percent of the customers chose the right-hand pair, 31 percent chose the mid-right pair, 17 percent chose the mid-left pair, and 12 percent chose the left-hand pair.

Word Maps

Another way for a magician to obtain specific verbal answers—always while exerting pressure and creating very determined contexts for psychological forcing—is to ask questions in such a way that the spectator, looking for the automatic answer that matches what the magician expects, is forced to respond without reflection. The magician is usually asking simple questions and anticipating answers that contain frequent or prototypical words stored in the semantic memory (such as "a canary is yellow"). Under these parameters, many magic tricks are based on asking for colors, numbers, geometric shapes, or objects; some expected answers even vary according to the genre or context in which the question is formulated.

The word maps, or "maps of familiar words," constructed by cognitive linguists are grouped by categories, and the words are classified by their frequency to determine the prototypical ones among them, which are representative of a certain category or class. Thus, for example, in the case of words designating animals, "cat" and "dog" are more frequent or representative than "pangolin." However, linguists know that the words deemed characteristic in a certain category vary a lot depending on the culture. If we ask a person in Spain to name a vegetable, they might answer "lettuce" or "tomato," but these will not necessarily be the first vegetables that come to mind if the same question is asked of someone in China.

Again, the success of the magician is directly proportional to his or her experience; experienced magicians are aware of these variations. Their greatest success, when it comes to asking questions of this type, is in formulating questions based on the cultures that they know best.

Be that as it may, because these are answers obtained automatically, immediately, and unconsciously, what we want to underscore is that spectators react by using the availability heuristic—that is, they give the first answer that comes to mind. Not only is the cultural factor very important, but so too are the conditionings (primings) for the evocation of certain choices that may have been previously induced. Alfred Binet has noted that, among chosen items, we always find the one that requires the least resistance, the least effort on the part of the spectator, and some magicians have acquired a great deal of experience in inducing these choices.[23]

The most commonly used forcings in magic are:

- Colors. The forcing of colors posits that red is usually the first color mentioned spontaneously, but that, if more time is given, it will be blue.
- Numbers. The forcing of numbers, also very common, stipulates that when the magician asks a spectator to choose a number "between five and ten" (this phrasing is better than "from five to ten"), or between one and ten, the answer will most likely be seven—a very high probability that has been exploited in magic for centuries.[24] (As Dani DaOrtiz points out, the decisive factor in a successful forcing is not so much the question itself but the way in which it is asked.[25] In the next section, we examine the importance of how questions are framed.)
- Suits. The forcing of suits in a deck carries a high likelihood that hearts will be chosen, although of course this probability must be treated with precaution.

The great magician Dai Vernon, nicknamed "the Professor," was a great modernizer of magic, and his influence was enormous on magicians throughout the twentieth century. Vernon performed a trick called "The Five-Card Mental Force"—for which, by the way, he had no plan B. It consisted of exposing on a table the following five cards, faceup and from left to right: the king of hearts, the seven of clubs, the ace of diamonds, the four of hearts, and the nine of diamonds. Then he invited the spectator to mentally choose a card, but he asked them to take their time, not to choose hastily. Then, after collecting the cards and shuffling them a little, he would leave only one of them facedown on the table—specifically, the four of hearts, because, according to Vernon, that was the card that was always chosen. Speaking of an American audience, Vernon explained that the king of hearts and the ace of diamonds were too obvious, the seven of clubs was an ugly card, and the nine of diamonds brought bad luck. In any case, we know that Vernon was priming for the four of hearts in his presentation of the trick, subtly emphasizing this card in his discourse over the others. In spite of its colorfulness, "The Five-Card Mental Force" is not a trick that has lasted because the result is risky, even if Vernon had great success with it.

Jay Olson, Alym Amlani, and Ronald Rensink have experimentally studied the perceptual characteristics of playing cards and observed that some are visually more accessible, others are better remembered, and some are chosen more often than others.[26] Among the most appreciated are the ace of hearts, the queen of hearts (mainly among men), and the king of hearts (mainly among women). In a survey cited elsewhere by Jay, the most popular suit was hearts, followed by diamonds. And the most spontaneously mentioned card was the queen of hearts, followed by the ace of spades, the seven of hearts, and the two of hearts.[27] No studies have reproduced these observations, although experiences reported in the world of card magic are consistent with them.

The Framing Effect in Magic

Magicians have also learned the importance of the framing effect—how the problem is presented or the question is asked. They know that presenting the same information in different ways can modify attitudes, perceptions, and choices. For instance, informing newly diagnosed cancer patients that they have a 10 percent chance of dying from surgery (negative frame) is different from explaining that they have a 90 percent chance of surviving (positive frame), and the patients' resulting decisions will be different.*

Let's take an example. Imagine that on the table we have a glass full of water and a glass that is empty. Then imagine that a participant in the

* In a study done on patients with lung cancer, patients had been given a choice between surgery and radiation. When informed that they had a 90 percent chance of surviving the surgery, 82 percent opted for it. But when told that they had a 10 percent chance of dying from the operation—a different way of expressing the same thing—then only 54 percent chose surgery. People who are faced with a life-or-death decision do not respond to the probabilities but rather to how they are described. See Edwards et al., "Presenting Risk Information."

experiment is asked to pour half of the water from the full glass into the other glass. If we then ask the participant to place the "half-empty" glass at the other end of the table, most participants will pick up the glass that was previously full. But when participants are asked to move the "half-full" glass, most of them will choose the previously empty glass.[28]

People usually react this way because the framing of a request conveys information in itself beyond what is explicitly being stated. People rarely consider this information consciously, but it still helps them interpret a situation intuitively and make decisions based on it.[29]

In magic, then, the framing effect is used as a tool to manipulate situations, particularly when it comes to inducing choices. Spectators may not think that the framing is important, but for the magician framing is not just a matter of asking the spectator to think, choose, indicate, point, touch, take, or pick up a card. Framing is asking the kind of question that will influence the trick's outcome.

Reflective Decisions

In contrast to instinctive, automatic decisions, reflective decisions are made in system 2, the system of conscious, rational processing. We have seen that, in reflective decisions, the brain does not react automatically but carries out a series of simulations, shuffling possible options based on experience. The reflexive decision process is alien to the heuristics that we have been discussing so far in this chapter. To make reflective decisions we evoke past scenarios, we simulate futures, and we end up making a certain decision, even though, as happens with instinctive decisions, reflective decisions can be modified by emotion.[30]

In any case, when analytical reasoning is imposed under system 2, judgments and decisions made under system 1 through shortcuts or simple rules can be improved and initial decisions can be modified. As we know, however, magicians avoid creating effects that require reflective decisions, because these are neither controllable nor adaptable to the necessarily short times of the tricks or to the quick, automatic, or instinctive answers and decisions sought by magicians.

Moreover, there is no room in reflective decisions for the disinformation techniques that magicians use to reinforce the appearance that a choice has been completely logical and free.

Reasoning in Hindsight in Magic

The disinformation introduced by magicians in some of their tricks can create situations similar to those that occur when we make "postdictive" arguments. This is reasoning that we consciously believe has led to a certain decision, but in reality is reasoning used a posteriori to justify the decision after the fact; we had already made the decision unconsciously.

Several studies by the Swedish researchers Petter Johansson and Lars Hall have shown that we humans can easily deceive ourselves and that we tell ourselves stories about our own decisions.[31] In 2005, these two researchers conducted experiments that led to an extraordinary observation.[32] They briefly showed volunteers several series of pairs of photos of people's faces. Then they asked them to choose the most attractive face from each pair shown. The experiment had a special phase: in some series, after the volunteers had made their choice, the researchers showed the chosen photo to the volunteers again and asked them to explain why they had chosen it. Then, unbeknownst to their experimental subjects, they sometimes changed one photo for another, using a sleight of hand so that the volunteers ended up looking at a face that they had not precisely chosen. The surprising thing about this experiment is that in most cases (75 percent) the subjects did not notice the change, and some offered reasons for their "choice." In one particularly striking case, the volunteer even justified his choice by saying that he liked the earrings worn by the girl in the photo when in fact the girl in the photo he had actually chosen was not wearing earrings.

Johansson and Hall call this phenomenon "choice blindness" because of its similarities to "change blindness." The novelty of their experimental paradigm—and what distinguishes it from previous experiments in situations of cognitive dissonance—is that not only were they able to change the outcome of their subjects' choices without their awareness of it, but they also recorded the subjects' reactions. When subjects explained (sometimes very elaborately) the choices they had never made, the authors were able to show them that what they were saying was not true.

These experiments thus offer a new methodology for investigating confabulation—the "stories" we fabricate, first for ourselves and then for others, to justify what we have done or said. (Let's also imagine the extent to which those who have participated in magic tricks may practice this storytelling.) But there is more. These experiments also show that confabulations can be so convincing that subjects eventually change their preferences and "consciously" choose the alternative they had previously rejected.

In another experiment, Hall and Johansson invited supermarket customers to choose between two types of jam.[33] Once they had decided, the volunteers were asked to try the variety they had chosen again and then explain why they had chosen it. But the jams had been switched, and the subjects ended up trying the option they had rejected. And again, only one-third of the volunteers detected the change, even when the flavors were as different as apple-cinnamon and grapefruit!

But what is really surprising is that these researchers were able to document the same phenomenon by simulating real online shopping situations involving cell phones, computers, and apartments, all of which involve much more reflection than choosing between brands of jam. In all cases, the experiment had the same results. The researchers obtained the same results even when questioning participants about aspects of life in which anyone would think that reflection plays a determining role: their political ideas (just before a general election in Sweden) and their moral positions (on universal social security, for instance, or the application of torture to those accused of terrorism).[34]

But Lars Hall and Petter Johansson's "trick" had yet another twist. When participants finished the tests, they were given a questionnaire asking what they thought of their own decision-making process. Among other questionnaire items, the subjects had to say how they thought they would feel if they had participated in an experiment that tricked them in this precise way. Of course, approximately 90 percent of the subjects said that they would never fall for such a trick—thus perfectly illustrating what Hall and Johansson call "choice blindness blindness." Like the volunteers in their experiments, not only are we blind to our decision-making process, but we also deny that such blindness can exist—that is, that such unnoticed manipulations can influence our decisions.[35]

A recent study has confirmed that these self-deceptions or false beliefs about our own past attitudes, as well as their association with confabulatory rationalizing (storytelling) processes, can lead to rather long-term changes in political attitudes.[36]

On the other hand, there are circumstances or conditions in which such manipulations are detected, although not necessarily consciously. A study carried out in Argentina on political decisions, based on the methodology of Hall and Johansson, replicated one of the main findings of the Swedish experiment and achieved a high level of deception because most of the participants did not realize that the researchers were manipulating their political opinions.[37] But the Argentine participants clearly expressed less confidence in the responses that had been manipulated. It was as if, despite not being aware of the manipulation, they had detected that something had gone "wrong" anyway. As a result, none of the Argentine study participants ultimately agreed to change their vote, unlike the Swedish participants, who did so relatively often. For the authors, the Argentines' resistance to change could be explained by an "unconscious detection" of (self-)deception—a phenomenon that they suggest is parallel to the unconscious metacognition of perceptual tasks.[38]

We wonder to what extent the spectators of a magic show may be able to invent things in hindsight about what they witnessed and about the "free" decisions they supposedly made if they were participants. We do not yet have the answers, but they will surely depend on the memorability of the events they witnessed.

PART III

The Results

CHAPTER 11

The Magic Experience and Its Audiences

> You experience magic as real and unreal at the same time. It's a very, very odd form, compelling, uneasy, and rich in irony. . . . A romantic novel can make you cry. A horror movie can make you shiver. A symphony can carry you away on an emotional storm; it can go straight to the heart or the feet. But magic goes straight to the brain; its essence is intellectual.
>
> —Teller, in Joseph Stromberg, "Teller Speaks on the Enduring Appeal of Magic" (2012)

Experiencing the Illusion of Impossibility

The magic experience is more than just doing tricks or fooling the audience. As Darwin Ortiz said, "Magic is not simply about deceiving. It's about creating an illusion . . . trickery is just a means to an end."[1] Gabi Pareras affirms that "the easy thing is to deceive, and the difficult thing is to achieve the magic of the impossible."[2] The true magic experience achieves the "illusion of impossibility"—the mystery that, according to Tamariz, can range in degree from the unknown to the "mental shock" that produces what is understood as truly impossible.[3]

The Emotions of the Magic Experience

Most emotion in magic is felt at the end of the trick, at its climax, and consists of the brief and transitory surprise the audience feels when faced with the impossible—the moment of the classic responses "Wow" and "No way!" As Arturo de Ascanio said, "Magic is leaving people with their mouths agape."[4] Surprise is the first emotion that appears in response to something unexpected or from a colossal conflict of expectations, such as the impossible effects in magic. This emotion can be accompanied by certain characteristic facial expressions, although they are not always expressed.

The surprise that spectators feel helps to focus their attention and facilitate memorization of the magic event, but it does not guarantee that their subsequent memories will be faithful to the lived reality. However, the opposite effect is also possible: if a magic show produces no surprise, the audience will probably later remember very little of what was seen and heard.

After the initial surprise at the unexpected or incomprehensible, what follows is usually fascination, confusion, laughter, or disbelief. For the philosopher Jason Leddington, the magic experience is aesthetically pleasing despite the various strong negative emotions sometimes provoked, such as feelings of vulnerability, uneasiness, loss of control, fear, confusion, or bewilderment.[5] The American magician Teller expresses it very eloquently: "Magic is a competitive, uncomfortable thing to watch."[6]

But if magic violates expectations, and the spectators do not know how magicians do it, why do they enjoy magic? Leddington examines the subject from a philosophical perspective. He describes the experience of magic as essentially "aporetic"—it leaves the spectator at a loss and perhaps intellectually uneasy, as when the plot leads to a dead end, a paradox, or an insurmountable logical difficulty. As Leddington explains, coexisting emotions such as wonder, surprise, and curiosity can inform and even partly transform coexisting feelings of bewilderment and loss of control. Such a transformation can lead, in turn, to a perception of these feelings as more pleasant, which explains not only how we can genuinely enjoy the magic that most perturbs and baffles us, but also why these performances can be so captivating, memorable, and emotionally powerful.[7]

The Unwilling Suspension of Disbelief

What, in short, is the essence of the magic experience?

The idea that the audience "suspends disbelief" when viewing a magic show is widespread. This notion explicitly puts a magic show on the same footing as any theatrical or cinematic fiction in which the audience is temporarily summoned to postpone their judgment on the impossibility of what they are seeing or witnessing.[8] Teller describes the difference in how "suspension of disbelief" is experienced in the two settings: "In typical theater, an actor holds up a stick, and you make believe it's a sword. In magic, that sword has to seem absolutely 100 percent real, even when it's 100 percent fake. It has to draw blood." Therefore, for Teller, theater is about the "willing suspension of disbelief," and magic is about the "*unwilling* suspension of disbelief."[9] Following his argument, "magic is a form of theater that depicts impossible events *as though* they were really happening."[10]

The experience of impossibility is directly related to the consistency of the presentation, as writers and artists know well. Aristotle, in chapter 24 of his *Poetics*, written in the fourth century BCE, stated that when writers are creating their work, they should prefer "plausible impossibilities" over "implausible possibilities."[11] For a *possible* situation to seem plausible, it need only be reasonably consistent. And for this to happen, it must provoke the minimum contrast in relation to our previous experience and knowledge of the world.

But what makes something *impossible* seem plausible?

Normally, the impossible is made plausible as part of a parallel story full of logic and verisimilitude. Samuel Coleridge, at the beginning of the nineteenth century, suggested that if a writer could give an appearance of reality to a fantastic story, readers would suspend their critical capacity and disregard the impossibility of the story.[12] In developing this concept in his 1939 essay "On Fairy-Stories," J.R.R. Tolkien went even further by arguing that the impossible story need not appear credible in the real world: it must only be consistent with a secondary reality that corresponds to the fictional world that has been created.[13] Readers or spectators remain immersed in that secondary reality and do not need to suspend their capacity for judgment. When they simply "unwillingly suspend disbelief," their sensations and enjoyment are enhanced. This is something that a magic trick achieves by creating a secondary reality expressed through the external life of the trick—which has its own consistency—and doing so without generating local contrasts that unduly capture the audience's attention before the end of the trick produces the denouement: the illusion of impossibility.

To experience magic, spectators must believe that they are witnessing something seemingly impossible but real enough that the experience produces vivid emotions, not dreams. The magician is not a character with superpowers who makes everything possible, but someone who makes others experience those impossible phenomena as real. For example, when we see Peter Pan flying in a movie, we know that what we are seeing is fiction. When we see the famous American magician David Copperfield flying onstage, however, we experience the flight as real, even though we know that it is impossible, since what we are seeing is not a dream.[14] In magic, the spectator does not have to imagine the impossible because the impossible is experienced as real. The impossible is not a fiction but an illusion, which is expressed, according to Teller, in this "unwilling suspension of disbelief." This is what Ascanio defined as a "magic atmosphere" in which "the audience doesn't suspect the mere existence of that cause [method], so that the series of effects always catches them totally off guard."[15]

The Magic Outcome as Cognitive Dissonance

Being offered a real (possible) experience of something impossible produces a cognitive dissonance that sometimes makes the audience emotionally uncomfortable. In fact, it has been observed that, when faced with a magic effect, the layperson's brain reacts by activating areas associated with the identification of conflicts.[16] However, unlike what happens when we "voluntarily" suspend disbelief while reading fiction, this dissonance consists, more specifically, of a conflict in our belief system; the conflict arises from our knowledge that it is simply impossible for a human being to fly unaided.[17]

The magnitude of this cognitive conflict depends on how impressive the magic effects were and on the wide range of individual variability in the sensibilities of the spectators. The degree of cognitive conflict depends, then, on what people consider impossible (beyond what the effects might be), and that is why people who have different beliefs or are from different cultures can react differently to the same effect.[18]

For the magician Darwin Ortiz, the magic act produces an emotional belief without the capacity for intellectual confirmation—along the lines of "I don't believe in ghosts, but I am afraid of them." The impossibility of magic is a conclusion reached by the viewer through a process of elimination, because no alternatives are offered. The main task of the magician is therefore to get the audience to rule out any kind of possible cause, thus canceling all possible explanations.[19] Thus, if the natural and immediate response of the audience to a magic illusion is to seek solutions to minimize such cognitive dissonance, the magician's job is to thwart these attempts.

As Simon Aronson stresses, "There is a world of difference between a spectator's not knowing how something's done versus his knowing that it can't be done."[20] The key to what will finally be the best "magic experience" is found in the construction of the trick and its presentation. The illusion of impossibility will not be the mere consequence of distractions or specific misperceptions, but the result of the whole structure and presentation of the magic act.

Wally Smith and his colleagues focus the discussion appropriately by defining the logical nature of impossibility as "an unresolvable contradiction between a perception-supported belief about a situation and a memory-supported expectation."[21] Scientists have begun to determine the neuronal correlates of this experience of the impossible. For example, when magicians shatter expectations, electroencephalogram (EEG) recordings have shown responses typical of the presentation of infrequent stimuli in standard laboratory paradigms, but with a delay that can be attributed to the fact that subjects needed more time to process the impossible magic outcome.[22]

In chapter 7, we covered the predictive role of the brain, without which no assumptions can be made and we are unable to interpret what we perceive. The brain functions in such a way that it is constantly correcting errors related to the hypotheses it is developing. According to Jakob Hohwy, we use conscious perception to minimize prediction errors, and through attention we control what is observed in order to optimize the accuracy of our expectations.[23] However, when a magic trick culminates in an impossible outcome, the prediction error is so great that our brains are unable to correct it and cannot deal with it. That is when the "magic experience"—the "No way!" reaction—occurs.

To our minds, impossibility is a probabilistic phenomenon, not a binary one, and magicians strive to get us to experience its fullest expression.[24] This experience includes intellectual and emotional reactions, and it takes place even when the spectator is fully aware that the magic effect is based on certain methods and techniques and that there is nothing supernatural about it.

The Validity of the Illusion of Impossibility

There has never been an exhaustive controlled study on the nature of the feelings generated by witnessing a magic effect.[25] The time is now ripe for such a study, given that, according to the British magician and psychologist Peter Lamont, we live in a time when audiences at magic shows not only know what they are going to see but also are no longer surprised in the same way that spectators at the end of the nineteenth century were. For Lamont, today's classic magic shows no longer produce a "surprise" due to the violation of natural laws; modern spectators experience only a sense of "astonishment" or "amazement"—a strong reaction to something exceptionally new or unexpected.

What has prevailed over the centuries is reaction to the *unexpected,* whether or not the reaction to the impossible is not so valuable today.[26] In any case, in the absence of a previous operative definition, the concept of astonishment and amazement can be synonymous with the concept of surprise, so this debate cannot go beyond a fundamentally semantic discussion.

In the survey by Joshua Jay mentioned in chapter 8, most participants stated that what they liked most in a magic show was the surprise, and what they disliked most was repetition or well-known classic tricks, unless they included unexpected effects. This survey confirmed that what audiences mostly like, in short, is "not knowing what will happen next." This is a paradoxical expectation because surprise, by definition, arises only unexpectedly; hence, in magic, audiences expect the unexpected.[27]

Magic and Superpowers

When people are asked what kind of magic effects they like best, mentalist effects get the most votes. Compared to audiences for classic magic shows, many more people willing to participate in mentalism shows.[28]

Mentalism is a branch of illusionism characterized by effects based on the supposed mental powers of the artist, such as telepathy, telekinesis, clairvoyance, and precognition. In mentalism, the illusion of impossibility does not occur in the same way because the personality of the mentalist often overlaps with the effect itself. Also, although at present nobody doubts that the magician uses tricks, the world of mentalism is more rife with ambiguity, and those who are devoted to it are usually reluctant to confess expressly that they use tricks for their show. Thus, there are still spectators who come to these shows convinced of the supernatural powers of the artist. Not surprisingly, mentalism techniques are also used, with less honest intentions, by fortune-tellers, seers, tarotists, psychics, spiritualists, and other kinds of tricksters.

But why is it that, in every show, there are some people who feel the urge to seek explanations of the occult or paranormal, despite magicians' open claims that their effects are based on magic tricks? This is a fact with deep cultural roots, and an in-depth examination of it is beyond the scope of this book.[29] Suffice to say here that many people have this predisposition, and it does not seem to be correctable no matter how clearly and eloquently the artist explains his methods.

Spectators who experience abnormal or improbable events and believe that they are in the presence of people with psychic powers can even change their beliefs and show cognitive biases that they did not have before. In one case, a group of students knew a priori that they were going to attend the show of a mentalist whose effects were based on magic tricks. Demonstrating how easy it can be to manipulate credulous people, some of the students were convinced after seeing the show, which had included a supposed conversation with a deceased friend of one of the spectators, that the artist possessed "psychic powers."[30]

We share the concern that magicians who embellish the story of their tricks with false scientific explanations are doing science a disservice, because such explanations reinforces beliefs in pseudoscience.[31] We also believe that at the present time, when pressure groups that specialize in spreading hoaxes and all kinds of falsehoods through social networks are proliferating, it is unethical for some magicians, especially mentalists, to engage in disinformation and contribute to the perpetuation of scientific falsehoods.

Magic in the Twenty-First Century

If scientific and technological advances and innovations force us to continually rethink the concept of what is "possible," it is logical to ask how magic in general is valued by present-day audiences.

Surely the vision of the world in the late nineteenth and early twentieth centuries was more conducive to certain magic tricks; today, given existing technological resources, spectators have many more tools at hand that help them decipher magic effects. Even so, many magicians achieve new effects using cell phones, tablets, drones, and so on, and achieve results that are just as effective as in the past.

Today we accept constant innovation and its products without much resistance, whereas a century ago people were more credulous about the supernatural, and any new technology had a great social impact. Today, by contrast, we are more educated, and we live with many more gadgets than ever before. Even if we have no clue how they work, they do not have a great effect on us.

But even in the twenty-first century, surprisingly enough, many magic effects that were performed a couple of centuries ago remain successful. The fact that spectators are still capable of being amazed by magic only reinforces the unique validity of this art. It is thus worth emphasizing that magic interferes with the automatic functioning of our brains, giving rise to a type of disruption that resists the passage of time and the evolution of our society.

Magic is a singular art capable of provoking the marvelous experience of the impossible with no need for audiences to believe that what they observe is real. And to this day, this complicity of the spectators—who attend magic shows knowing perfectly well that the magician is an artist who secretly uses special materials and methods—still permits them to enjoy the illusion of impossibility.

When the situation of impossibility is experienced, when expectations are violated, specific areas of the brain related to problem-solving and conflict identification are activated.[32] Amory Danek and her coauthors have postulated, through magnetic resonance imaging studies, that these areas reflect the conflicted reaction to the illusion of impossibility, given that they are not activated in magicians when they watch their own trick performed.[33] Not only are magicians unable to perform magic on themselves, but the audience's participation is essential for the trick to work. The magician and the audience form an inseparable pair. This is not the case in other arts, such as music, which allows listeners their own interpretations, or in other activities, such as sports, which can be played behind closed doors.

Recently, attempts have been made to reproduce magic tricks, with their impossible outcomes, using computers. Using artificial intelligence, researchers have applied mathematical techniques to humans' physical and cognitive limitations to create and optimize potentially new and original methods of doing magic. Although researchers recognize that these systems are far from being autonomous and still need a lot of human intervention, this interesting approach to magic from the perspective of AI deserves to be followed closely.[34]

Is Live Magic in Front of Spectators the Best Magic?

The credibility of a magic trick with an impossible outcome is probably very dependent on the distance between the magic and the spectators. Prerecorded magic acts (on television or in any other audiovisual media) can supposedly be loaded with special effects, but they produce a different kind of magic experience that hardly surprises with the same quality, disbelief, and intensity as magic acts experienced in person.

Even when magic is live, distance is a crucial factor for the illusion of impossibility. For example, close-up magic usually has a greater impact than stage magic by the sheer power of its proximity. And of course, the experience of live magic is stronger than when televised.

To achieve the best possible experience, the magic trick must be done live, but is there any relationship between audience distance and impact? Generally speaking, the shorter the distance the greater the impact and, in turn, the more difficult it will be for spectators to think about the methods used by the magician. The opposite is probably also true, in the sense that, at a greater distance, to achieve an impact comparable to that achieved in some close-up magic acts, the effect must be on a grand scale, as when the magician David Copperfield made the Statue of Liberty disappear.

Magic can also vary in intensity. In Joshua Jay's survey, the scale of the effects was observed to be greater than the distance, and his survey participants said that they would prefer the disappearance of a helicopter from a stage than the disappearance of a coin from under their noses.[35] Thus, for most respondents, scale seems to be more important than proximity.

When asked what is the "stronger" effect, making something disappear or changing or producing something, most respondents said that the strongest effects are changes in something.[36] Again, and not surprisingly, magic follows many of the cognitive strategies that we humans use to understand the world.[37]

Magic Audiences

To speak of the audience is to speak of an essential component in the magic experience, because without an audience there is no magic.

Magic is essentially participatory. Unlike theater, there is generally no "fourth wall" in magic. Instead, the magician establishes a direct relationship with the audience members, considers them, addresses them, and invites them to participate.

For the first time in history, that "fourth wall" was reintroduced when live magic shows were banned—breaking the continuity between magicians and audiences—during the Covid-19 pandemic. Needing to reinvent themselves, magicians created a new online paradigm and interactive situations with the participation of the audience. Some of them even offered an unprecedented advance registration a few days before the show so that they could ship to audience members a box of items that they would subsequently manipulate from the other side of the screen.

This virtual proximity created by interactive events was an intermediary step between canned magic and live magic. However, it was not enough: the screen imposes a distance that is not only physical but also conceptual. Everything is possible across the screen. And if everything is possible, magic simply does not exist. It becomes, at best, another visual art form and, at worst, another visual illusion.

In addition to varying by situational contexts, magic also varies depending on the audience to which it is directed. In this sense, there are at least three types of magic: magic for laypeople, magic for children, and magic for magicians. To ask a separate and very interesting question, in the future could there also be a specific magic for artificial intelligence?[38]

The first category, magic for laypeople, covers quite a range: amateurs who genuinely enjoy magic and probably make up the majority; a minority who are provoked by the magician and have a low tolerance for the tricks; and finally, some people (we do not know how many) who seem to feel nothing when watching a magic show.

As we already pointed out in chapter 5, magic takes advantage of the context. The reactions, experiences, and memories of the audience probably differ if we compare the experience of magic tricks performed, for example, impromptu with friends or in a restaurant with the experience we have when attending a show. There may also be differences between the occasional spectator and the amateur who accumulates experience over time, especially when watching the classic tricks. We wonder to what extent this accumulated experience conditions the amateur to repeatedly see certain classic games that have been done practically the same way for two hundred years.

Many amateur spectators admit to preferring certain magicians; for them, it is not the tricks that attract but the magician's character, charisma, style, and poise in front of the audience. Miguel Ángel Gea writes:

> The audience always wants to fall in love with whoever is on stage, he amazes them because they couldn't perform the same feats as the actor but would love to do so. This infatuation serves as a cover and even allows the spectators to forgive a failure. This infatuation allows them to let themselves go, be overtaken by what's proposed to them, and be dragged into the external life of the trick by the actor's charisma, with the result that the trick is even more mysterious.[39]

It is also said that, as noted by Max Dessoir in 1893, educated spectators are a more ideal audience for magic and enjoy it more than less educated spectators. This is surely because educated people are more likely to discard apparently superfluous information and to infer and anticipate more effectively. Educated spectators are more likely to react to the contradiction of the effect because they are also more aware of the impossibility created by the magician. Thus, for John Mulholland, "magic is designed to fool the brains, not the eyes. This means that the best audience for magicians are persons of intelligence."[40]

Are there spectators who can more easily deduce the methods behind such tricks? In chapter 9, we discussed how viewers can "notice" (unconsciously) the methods in certain tricks, even when the execution of the trick is highly skilled. Do some spectators have greater sensitivity or facility in learning the subtleties of the methods? Magic tricks are often not performed well enough. Some artists put in insufficient practice and keep the kind of very low profile that is highly unlikely in other arts—for example, in music. It is also known that spectators can be very forgiving: sometimes they see things they should not have seen, but they allow themselves to be carried away by the show anyway, even if it is not up to par.

This benevolence of spectators who have detected the magician's methods would be unacceptable in other fields—for instance, think of a musician playing off-key. Alfred Binet wrote, way back in 1894, that when magicians wanted to force a card, they approached a woman, because it was assumed that women would always be much more reticent than men if they noticed something.[41] Thus, we see that magicians have always counted on indulgence. This pact of silence with the audience is an aspect of magic that has been little explored systematically to yield new knowledge about how magic is processed.

Magic for Children

Figure 11.1 shows three rows of drawings of neurons. One row was drawn by biomedical undergraduates, one was drawn by neuroscience doctoral students, and one was drawn by neuroscientists.[42]

Which row was drawn by each of these three groups?

Here's the answer: the top row was drawn by undergraduates, the middle row was drawn by doctoral students, and the bottom row was drawn by neuroscientists.

Let's look at the differences between the top and bottom rows. Undergraduates do not have a thorough knowledge of what a neuron looks like, so they draw it just as it appears in a reference textbook; they might even add labels to the different parts, in case it is unclear what they are referring to. The undergraduates' drawings of the various neurons are very similar to one another.

What do we want to demonstrate with this example? That as we gain experience, we learn to separate the wheat from the chaff so as to be left with only what is relevant. Each drawing reflects an individual's learning history.

Established scientists made the drawings in the bottom row. These are schematic drawings with few details, reflecting what is essential for each individual. The reason for these differences between students and experts lies in the way humans conceptualize information. Over time we generate a codification system based on a commitment to minimal information. And through experience we learn to store only what is

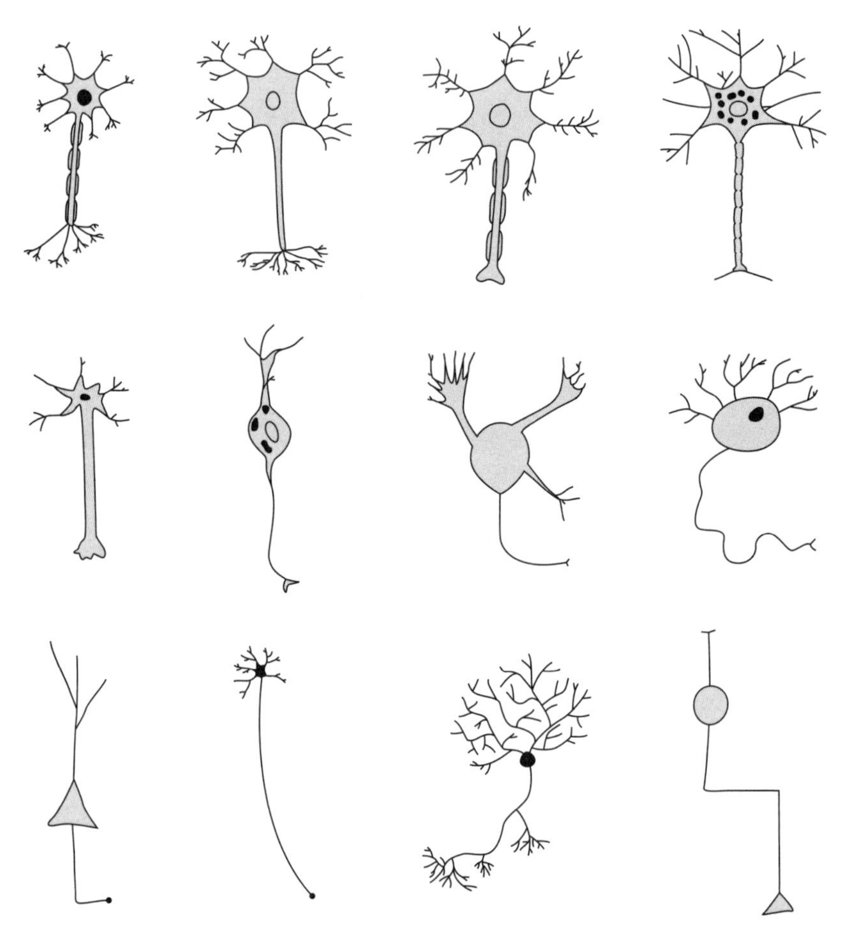

FIGURE 11.1. Drawings of neurons: The drawings are grouped into three rows, according to the age or experience of the people who drew them.

most relevant for the recognition of each concept and to eliminate details that we do not even process. Experts therefore discard much more information than students do, as the latter are still trying to grasp what is most relevant.

All this is directly related to a common understanding among professionals in the field of magic that magic for children must be different from magic for adults.

Children detect many details in the methods of magic tricks intended for adults because of the difference that we have just revealed between students and experts and also because of the way in which the human brain develops from birth. As our brains mature, they first develop the basic brain functions—the motor and sensory areas—and then the regions involved in spatial orientation, speech, and language. Structures involved in executive functions, such as the frontal lobes, develop later. Development of the prefrontal cortex is the most delayed, and the lateral prefrontal cortex matures last. It is precisely these underdeveloped brain structures in children that are associated with the filtering of ideas. In addition, the capacity for attention does not reach adult levels until the age of seven; it is still developing between the ages of four and six.

In short, the ability of the human brain to infer the physical properties of the world is shaped both by innate constraints and by past experiences that reflect statistical regularities in the world. That ability also evolves with age, so we select information that is useful to us differently at the beginning of our lives than we do in adulthood. Thus, children give much more importance to details and not so much to global concepts. Because they want to assimilate everything, they have a notoriously hard time focusing their attention. They end up gathering all kinds of superfluous information, and that is what makes children recognize the methods being used by magicians in certain magic tricks.[43] Indeed, since magic tricks go against what we typically see in everyday life, they are the perfect tool to investigate the developmental acquisition of adult intuitions.[44]

While their brains are developing, children are more open-minded and naturally uninhibited. Thus, they usually see the actions of magicians as what they are, not as what we adults suppose them to be.[45] In addition, children are not shy but will say what they think; they are unlikely to make allowances the way adults do.

Based on all the preceding, not only must the language change in a magic show for children, but so too must the rhythm and the type of tricks. (And although we will not go into the nuanced differences among children according to age, we do want to acknowledge them.) To give a concrete example, children need for things to happen more frequently, preferably every fifteen or twenty seconds, because, as the Spanish magician Woody Aragón rightly states, "They say that children don't like card tricks, but I think that what they don't like is being bored." Children also do not like "super skills," according to Aragón, "because that's what they have parents for."[46]

In addition, children have a preference for the visual. To them, cards are no more than pieces of cardboard moving in and out of pockets or shifting in a deck. For children, the suits and values of cards are of no interest, and unlike adults, they are not impressed with divinations or predictions, let alone any experience of mentalism. Instead, children enjoy tricks that involve teleportation, transpositions, and disappearances.[47] They also like the "Miser's Dream" trick, color-changing handkerchiefs, the "Magic Coloring Book" trick, and the cut-and-restore rope trick.

Finally, children are fascinated by a coin coming out of their ear, but they have little interest in a coin changing from copper to silver or the back of a deck of cards changing color.

When Magic Provokes the Spectators

When they see a magic trick, some people feel so challenged to discover "how it was done" that they miss out on the pleasure of the magic effects. Luckily, magicians are well aware that there are spectators of all tastes and that some of them (fortunately just a few) do not tolerate magic very well.

Some of these spectators are hecklers who react negatively, sometimes even aggressively, and can try to sabotage the magician while the trick is being performed. Yet the worst thing that can happen to a magician when he invites a spectator to collaborate in a trick is to unintentionally choose a troublesome person. The worst reactions of this type are experienced when the audience includes only a few people in a small venue, such as a bar. Such spectators sometimes demand that the methods be revealed and pose unreasonable challenges to the magician.

Why do some people behave this way?

Many believe that these people cannot tolerate the admiration that magicians garner with their effects or the symbolic "power" of magicians who work wonders (and who do not reveal their secrets). Most likely, these people suffer from great jealousy at not being the center of attention for the evening. A recent attempt to psychologically characterize people who hate magic reveals that they are marked by lower openness to experience and a higher likelihood of socially aversive traits.[48]

The operative word here is "power" because, from primitive times, the attraction of magic is intimately related to the subconscious desire to dominate the uncontrollable world in which we live. As Teller wrote,

> In magic . . . there's an explosion of pain/pleasure when what you see collides with what you know. It's intense, though not altogether comfortable. Some people can't stand it. They hate knowing their senses have fed them incorrect information. To enjoy magic, you must like dissonance.[49]

There are also people who are not moved by magic at all, who are invited to see an effect and do not even get upset. Magic just does not appeal to them, in contrast to amateurs.

If "amusia," commonly known as tone deafness, is the inability to recognize musical tones and find them pleasant, perhaps we could use the term "amagia" to describe a person's inability to recognize magic effects and enjoy them. How prevalent might amagia be? Is there some neuroscientific reason for it, or is it cultural in nature? For us, amagia is an intriguing phenomenon that no one has ever studied; perhaps it is much more prevalent than we think given that, for some reason, magic is an art for the few.

Magic for Magicians

Finally, let's talk about a completely different world. Spectators who are themselves magicians are much more than fans. They are the great lovers of magic.

In fact, the passion involved in magic is such that there are magicians' magicians who will mainly or exclusively do magic only for other magicians. Magic tricks for magicians take advantage of the knowledge that these very special spectators already have. The tricks are generally subtler, more elaborate, and more complex. Moreover, the feelings experienced when magicians manage to deceive their own colleagues are also different. The tricks designed for an audience of magicians are often inappropriate for a lay audience. Beyond what is required to create the illusion of impossibility, magicians value the skills required and appreciate the uniqueness and complexity of the construction and methods used in these tricks.

This sometimes confuses many magicians and amateurs who, after becoming familiar with magic for magicians, lose the ability to select the best tricks for a lay audience. Magicians and some amateurs thus have the misfortune of losing their innocence—the capacity to feel the surprise and mystery that the illusion of impossibility evokes.

Seneca, the Hispano-Roman Stoic philosopher born in Cordova, Spain, in the first century BCE, insightfully expressed his lack of interest in discovering the secrets of street magicians and possibly losing the charm of magical illusion they created:

> Such quibbles are just as harmlessly deceptive as the juggler's cup and dice, in which it is the very trickery that pleases me. But show me how the trick is done, and I have lost my interest therein.[50]

It has been proven that spectators of a magic trick can perfectly well distinguish between what they see and what they think is happening.[51] In chapter 8, we discussed the "aha!" effect: the moment when laypeople believe they have discovered how the method works, even if they are wrong, the illusion of impossibility vanishes with the assumption of knowing. The difference between performing for magicians and performing for laypeople is that, if magicians manage to deduce how 90 percent of a trick works but not the other 10 percent, they feel that the performer has managed to deceive them. However, if laypeople are able to find out or infer even 10 percent of what they see, even if they do not know how the remaining 90 percent works, they conclude that the trick has not deceived them and will surely go away without feeling the magic. This disappointment is even greater when an amateur enters the world of magic and discovers the overwhelming simplicity of many of its methods—a simplicity that succeeds by interfering with the automatic inferences that our unconscious brain is constantly making.

The Popularity of Magic

Magic was very popular in the late nineteenth and early twentieth centuries, but audiences then were very different from today's. It was a time when more people gave credence to paranormal phenomena and a smaller proportion of people were educated compared to today.

In short, these were times when magicians competed fiercely with mediums (at that time most mediums were women), who supposedly possessed supernatural abilities, including the power to connect spectators with their dead ancestors. At the end of the nineteenth century, "magic" thinking—in its esoteric sense—also began to coexist for the first time with scientific thinking.

Harry Houdini's relationships with spiritualists constituted a turning point in his life. Deeply affected by his mother's death while traveling in Europe, Houdini agreed to participate in a session with the wife of his friend Sir Arthur Conan Doyle. Using "automatic writing," the medium intended for the session to put Houdini in touch with his mother. Houdini's skepticism of spiritualists grew even stronger after this experience, however, because the text that the medium wrote was in English, while his mother, despite having lived more than fifty years in the United States, never understood, read, or wrote that language. In his own words, "I was willing to believe, I even wanted to believe." Houdini's progressive disillusionment pushed him into what was later a long and active career aimed at unmasking the fraud of spiritualists and practitioners of the paranormal.[52] This is why he advertised shows that would prove that spirits did not exist and in which he would expose the trickery of mediums (figure 11.2).

Throughout the twentieth century, the popularity of magic gradually declined, largely coinciding with the rise of cinema and then television as great instruments of mass entertainment. Magic became a minor entertainment, relegated to children's parties, extravaganzas, and variety

FIGURE 11.2. Poster for a Harry Houdini show in which he proclaims his criticism of mediums.

shows. Near the end of the twentieth century, however, the landscape changed again: magic has become increasingly popular, probably because of the proliferation of television magic shows, outside theaters, and street performances, and also owing to the influence of social media among young people.

Although interest in magic has revived, the art has carried over some old practices. Nevertheless, what has changed the most about magic is the diminished secrecy around its practices; consequently, its circles and associations have become less impenetrable. Whereas knowledge of magic was once a privilege, today, fortunately, it has become a right, and anyone who wants to learn about the art can do so. Literature specializing in magic is accessible to any amateur. And social networks make everything available to those willing to look for it.

Compared with other performing arts, magic still does not draw a big audience, and it still tends to be associated with entertainment for children, although not as much as in the past. Nevertheless, most magicians, whether professional or amateur, confess that magic is a true passion for them. Many also admit that they were hooked when they were children, which raises some questions: Is it perhaps the possibility of having "supernatural powers" when one is a child that awakens the interest in magic? And if so, does this have anything to do with the gender roles that are reinforced later?

It has also been claimed that women are not as interested in magic as men are. Is there any basis for this claim? Does the "power" that emanates from traditional magic tricks have something to do symbolically with male roles? In analyzing this gender asymmetry from a social point of view, Peter Nardi has highlighted not only the importance of gender roles but also the consequences of the aggressive rapport often established between the magician and his audience.[53] How often does the artist who offers a "look at me" type of magic, with an essentially egocentric *performance*, contribute to this asymmetry? Is this style of performance directly related to the scarce number of women magicians today as well as to their traditional supporting role in magic shows?

In any case, more women are becoming magicians today. The proportion of women magicians is greater, and, fortunately, more and more men are playing the role of partner for both male and female magicians.

Why, then, as magic adapts to the modern world, has it not become as universally popular as cinema? Why does it have fewer spectators than theater or the circus?

For the philosopher Warren Steinkraus, magic is a minor art form because it offers "minimal emotional consequences"; the climactic impact is ephemeral and rarely moving in any lasting way.[54] But in fact, magic is capable of provoking very strong emotional sensations, as some readers must certainly have experienced at some point. Perhaps the problem is that magic poses unlikely situations for which our brains are not prepared because they are completely foreign to us or at least outside our daily experience. In contrast, cinema and theater present us with plausible scenarios that we may have experienced ourselves or moments that make our imagination soar in a realistic way, acting on brain circuits that have evolved for our survival.

This could explain not only the relatively low popularity of magic but also the very existence of amagia.

The truth is that we still do not know much about magic as an art for human communication and entertainment, or about its sociological and neurological bases. The scientific research on magic is not very extensive, but as we will see in the next chapter, it has already begun to bear fruit. We can now anticipate a future in which, rather than simply using scientific methodologies to study magic, the magic techniques themselves will become an essential tool for the study of the human mind and sociology.

CHAPTER 12

Wrapping Up

Scientific Research and Magic

We presume that everyone will agree to the recognition of magic as an art. As a matter of fact, magic embodies both art and science.

—Nevil Maskelyne and David Devant, *Our Magic* (1911)

The Science of Magic

Often in scientific research, to make observations as objective as possible, experimental designs have reached extremes that are far removed from reality—as when the experiment is carried out in a laboratory, with the subjects isolated, watching magic recorded on a computer and unable to move while the researchers monitor their reactions. In a contrary trend, current experimental models have become more ecological, with the magician performing live before an audience. This model is closer to reality because, after all, the goal is to make the results as representative and generalizable as possible, but it has a steep price: experiments conducted in especially realistic conditions are much more complex and expensive.

In the context of classic, laboratory-based cognitive experiments, a hypothesis is generally considered confirmed when the results are 60 percent positive. In a live magic session, however, the magician cannot help but hope that 100 percent of the spectators will experience the illusion of impossibility, without exceptions! Ideally, this would be the scenario in which all magic experiments are done, and the result to which all scientific experiments on magic should aspire.

Magic tricks can be the subject or the object of scientific research. In other words, in addition to being the focus of research itself, magic can be used as a privileged toolbox for the scientific understanding of how the brain works. Neuroscience can learn from the experience that magicians have accumulated. Magic tricks can be magnificent resources for studying effects and mechanisms related to brain functioning in order to understand how people behave and how audiences are managed. The collective behavior of audiences in manipulated environments is a very topical subject, but it requires a much deeper knowledge than is currently available.

Is There a Scientific History Related to Magic?

Many magic tricks and almost all key methods in magic have been performed and practiced for hundreds of years. The methods used in magic were discovered to be effective in a totally empirical way, based on trial and error, by those in the magic community, who had little knowledge of the cognitive processes that we cover in this book.

Throughout the evolution of magic, renowned magicians have made important theoretical contributions to the fundamentals of magic tricks, such as how best to construct and present them. From the end of the nineteenth century to the middle of the twentieth, the work of Robert-Houdin, Dessoir, and, later on, Maskelyne and Devant stands out.[1] In this book, we have also cited the later works of Fitzkee and Sharpe, as well as the extensive theoretical contributions of other magicians throughout the twentieth century, including Ascanio, Randal, Bruno, Tamariz, Ortiz, and many others. For a closer look at these authors and their theoretical contributions, we refer the reader to *Magic in Mind*, a very interesting compilation edited by Joshua Jay and published in 2013.[2]

One of the fathers of modern magic was the French magician Jean Eugène Robert-Houdin (1805–1871). He took magic off the street and into the theater. He also made countless contributions, some of which are still reproduced onstage. In Robert-Houdin's treatises on magic tricks, texts that he began with some "general principles of prestidigitation,"[3] he dealt with such diverse aspects as the show's structure, the demeanor that magicians should exhibit, and their appearance—he recommended wearing a tailcoat. Although Robert-Houdin was not the first magician to wear one, his sartorial choice became a model.

It was precisely at the end of the nineteenth century, at a time when magic was still linked to religious and supposedly supernatural experiences, that the formal interest in the science behind magic arose. Many magicians, like the nineteenth-century French magician Richard, affirmed that they should not claim supernatural powers for themselves.[4] In those days, and for the first time, some famous magicians collaborated with scientists by revealing their tricks. We cannot overemphasize the value of this collaboration between professionals in two fields with such opposing methodologies—one of them based on deception and secrets and the other on transparency and open communication. This interest in collaboration was triggered by the increasing success of mediums, who flaunted their supposed supernatural powers to stage spectacles that competed directly with illusionism.

Alfred Binet (1857–1911), a psychologist at the Sorbonne, was a notable pioneer in the study of the foundations of magic; it was he who introduced the Binet-Simon scale to determine IQ. Binet studied prominent magicians of his time such as Arnould, Dicksonn, Méliès, Pierre, and Raynaly.[5] In 1894, Binet published "La psychologie de la prestidigitation," in which he revealed early on that magicians controlled the audience's gaze and attention. Binet also proposed that "passive illusions," from the perspective of magic, could be broken down into "positive illusions" (seeing what does not exist) and "negative illusions" (looking but not seeing). These concepts were based on James Sully's archaic distinction between active illusions, such as hallucinations, and passive ones, which everyone could experience.[6] Concepts that we now know as open or covert deviations of attention were clearly already empirically grasped.

A contemporary of Binet's, the German philosopher and psychologist Max Dessoir (1867–1947), upon discovering the importance and richness of psychological techniques in magic, argued that psychologists had much to learn from magicians.[7]

The outstanding work of the American psychologists Joseph Jastrow (1863–1944) and Norman Triplett (1861–1934) followed Dessoir's line of inquiry.[8] Jastrow worked with the two great magicians and rivals Alexander Hermann and Harry Kellar and applied psychophysiological techniques to test whether they had above-average abilities.[9] In a version of his doctoral thesis published in the *American Journal of Psychology* in 1900, Triplett proposed a very extensive taxonomy of magic tricks, outlining the methods used by magicians. The categories he proposed were tricks involving scientific principles, tricks involving superior mental abilities (such as mathematics), and tricks depending on the use of gimmicks, sheer manual dexterity, or fixed mental habits in the audience.[10]

Scholars in the late nineteenth and early twentieth centuries identified many of what have always been the great principles of magic, as explored in this book. Binet, for example, highlighted the importance of optical illusions in magic, spoke of screens that prevented perceptual reconstruction (by amodal perception), and described exogenous capture with an overt diversion of attention (what he called the ABCs of the craft) and covert diversion (looking but not seeing). He also provided a good example of divided attention, warned that the illusion of magic tricks was lost with repetition, and, as noted in chapter 4, experimented with chronophotography to identify those tricks performed so fast that they are undetectable by the eye.

As for the principle of non-repetition, it is interesting to read the advice of the British Charles De Vere to include different methods in the same trick if the same effects are to be repeated, a practice still followed today.[11]

The nineteenth-century French magician Richard focused most of the "preliminary instructions" for his tricks on advice based on the principles of nonrepetition and non-anticipation. Even back then, he encouraged magicians to be inventive and to strive to rejuvenate old tricks, "if not in substance, in form."[12]

Max Dessoir, besides anticipating concepts that would later be published by Binet (open deviation, divided attention), highlighted the illusion of impossibility, the necessary naturalness of the magician to avoid contrast, and the magician's need not to reveal where he is going with his tricks. Furthermore, Dessoir, in a time when nobody talked about what memory was, described techniques of disinformation and the elaboration of false clues while emphasizing that not everything that happens is going to be remembered. Dessoir felt that the word "prestidigitation" was not the best term for magic: it is not that the spectator is surprised by the marvelous speed used, he believed, but that the magician's success stems from *ars artem celandi*, art concealing art. According to Dessoir, magic is the art of disconcerting the spectators to such an extent that they are unable to suggest any solution to the wonders they have witnessed, and they go home accepting the explanations as conclusive, with a lingering feeling of having briefly lived in a world of wonder.

As mentioned in chapter 11, scientific interest in magic faded in the twentieth century, coinciding with a gradual decline in magic shows, largely because of the great expansion of cinema and television. More recently, the world of science has renewed its interest in magic and its mechanisms. But why did magic tricks receive little attention from the neuroscientific community for more than a century? An editorial by the renowned neuroscientist Richard Gregory, published in 1982 in the journal *Perception*, asked this very question. He remarked that the topic goes unmentioned in Hermann Helmholtz's treatise on physiology, which is considered the "bible of perception."[13] Was this lack of interest related perhaps to the low visibility of magic shows and the smaller audiences during that period?

Although we have no clear answer to that question, we are confident that the accrued experience of magicians and the wealth of knowledge to be derived from magic tricks can enrich human knowledge about how the brain works. But proving this point requires work.

The current work of cognitive science to understand how magic works is led by, among others, Gustav Kuhn, the Swiss psychologist and magician. He has published several studies on the psychology behind magic tricks and theorized about the potential contributions of magic principles to the advancement of cognitive sciences. Kuhn was the lead author of one of the first experimental studies on divided attention and the perception of a magic trick.[14] In 2008, he published an article in which he argued for the value of developing a "science of magic."[15] Subsequently, Kuhn wrote other articles in which he elaborated on his theoretical proposals.[16] One of these articles detailed an exhaustive taxonomy of tricks, always from a psychological perspective (see the section on misdirection in chapter 6).[17] More recently, Kuhn has presented a plan for comprehensively and systematically tackling the study of the effects and methods of magic tricks from the same perspective.[18] Other authors have followed a similar path, also from a psychological perspective.[19]

Simultaneously, in 2008, a team led by Susana Martinez-Conde and Stephen Macknik proposed the need for a more causal approach to magic tricks, one based on their neurobiological underpinnings.[20] These neuroscientists explored some neuronal processes involved in magic tricks, in consultation with well-known magicians after a joint meeting in 2007.* They later became extraordinarily popular after the publication of *Sleights of Mind*, a best-seller about magic and neuroscience.[21] This notoriety, in turn, unleashed greater interest in both neuroscience and magic.

Peter Lamont, a psychologist and magician, objected to both Kuhn's and Macknik and Martinez-Conde's proposals.[22] Arguing that the construction of a "science of magic" was an impossible goal to achieve (itself an illusion, if you will), he felt that the most realistic goal was to use magic tricks as tools to study cognition and other brain functions.[23] Lamont himself acknowledged, however, that "there is currently no reason to believe that such processes are any different from those that have already been identified, or might be identified, in terms of attention, awareness, persuasion, deception, belief, and so forth."[24]

To show that magic tricks are based on common mechanisms already studied in neuroscience has been precisely the purpose of this book. We have presented the cognitive processes behind the effectiveness of the diverse methods and techniques used to achieve the illusion of impossibility. In doing so, we have provided a fresh look that is focused not on

* The meeting was the "Magic of Consciousness" symposium at the eleventh annual meeting of the Association for the Scientific Study of Consciousness, Las Vegas, United States, June 22–25, 2007.

the phenomena of magic, which is what characterizes the proposals of Kuhn's group, but on how those techniques and magic effects could be theorized and systematized to be used as experimental tools and guides in cognitive neuroscience.[25]

During this tour, we have cited some of the most relevant, contemporary research-based literature relating to the impossible illusion provoked by magic tricks.[26] A partial list of this literature appeared in a special issue of the open-access journal *Frontiers in Psychology*, entitled "The Psychology of Magic and the Magic of Psychology."[27] In the aftermath, the Science of Magic Association (SOMA) was created.[28] SOMA organized its first two congresses in 2016 and 2017 at Goldsmiths, University of London, and in 2019 it held its first conference in the United States, in Chicago. SOMA "promotes rigorous research directed toward understanding the nature, function, and underlying mechanisms of magic."[29]

How Could Magic Contribute to Neuroscience?

Exploration of the relationship between science and magic so far has been mostly one-way, based on the work of a few scientists doing research to understand how magic works. We believe that the field should take a different but complementary route, reversing the direction of explanatory work. Instead of using the brain and behavioral sciences to study magic, we want to highlight the opportunity to use magic to study the brain and behavior.

The different cognitive processes involved in the magic tricks we have analyzed here suggest unexplored routes of research in which magic could contribute to a better understanding of human cognition. For example, we do not know the basis of the cognitive dissonance that leads to the illusion of impossibility, or its characteristics, degrees, types, and neural correlates, or whether these differ from those of deception and surprise, among other effects. Magic is experienced very differently depending on the context in which it is performed, the particular cultural background of the spectators, and the wide range of individual reactions to magic. As we have already explained, magic is also very different depending on whether it is directed at adults, children, or other magicians. Magicians rarely experience the illusion of impossibility, although they enjoy the technical skills and conceptual innovations of their peers. In addition, the reasons why magic does not draw huge crowds (compared to movies or serials) and why some people dislike magic tricks are, surprisingly, still open questions.

Beyond illuminating the questions that arise when we consider cognition through the lens of illusionism, magic can help us with the general challenge of developing a real-world neuroscience. Neuroscience and behavioral sciences have thrived in laboratory conditions, but do the processes measured in such artificial contexts accurately capture phenomena that occur in the real world? In fact, the very effectiveness of magic in real situations, compared to the lesser success of laboratory studies, invites a reexamination of many cognitive paradigms that have eluded an ecological approach. Most of the few existing research studies have been carried out under extreme laboratory conditions—that is, magic tricks are performed on video, often by less experienced magicians, with very few participants and without a representative audience. What we propose here is that, for several reasons, scientists should essentially adopt the role of the magician.

First, a simple magic trick can integrate many different cognitive processes at once. Alternatively, magic effects can be designed that target a particular cognitive process, allowing the experimental dissection of cognition into possibly more natural tasks than those normally explored in artificial laboratory environments. Second, while magicians always take the context into account when designing and presenting their effects, it is the other way around in laboratory experiments. In the lab, experiments attempt to eliminate any contextual information not directly related to the concrete process under study, but doing so takes us far away from reality. Moreover, ensuring the absence of contextual information does not prevent participants from subjectively taking the context into account, adding a layer of confusion not considered by the scientist.

Third, magicians do not base the success of their tricks on statistical measures that blur the individual in favor of a nonexistent average spectator. They address each and every attendee and often with complete success, as we have discussed. Fourth, magicians perform their cognitive manipulations in real time, in direct contact with the audience, and in a single trial, since they cannot afford to repeat the trick if it fails. Finally, beyond individual behavior, social dynamics are a significant component of magic, as the individual cognitive processes of each spectator are combined with the group dynamics that emerge spontaneously from signals such as applause or laughter.

Following the analogy between science and magic also reveals unexplored conceptual territory. Consider the two realities that coexist in magic tricks, internal and external life. It is not unreasonable for magicians to conceive and perform their magic tricks in a manner comparable to how nature operates. Therefore, scientists may be compared to spectators of a magic trick and impelled to discover how nature works from their own particular and limited point of view. We would even go so far as to speculate that some of the mysteries of how the brain works are discernible in the distinction between the external and internal realities of any magic trick.[30]

In short, for neuroscience to benefit from magic and its refined methodology, overwhelming successes, and unique perspective, it is essential that the joint work between magicians and scientists continue. This is an area of interdisciplinary research with a long journey ahead. In fact, from the late nineteenth century to the present, fewer than one hundred experimental research articles (not including reviews and editorials) in which magic tricks have been used as a source or as a research objective have been published.[31] That is a very low figure from all points of view, and it would not be an exaggeration to say that, compared with the state of research in any other scientific field of study, when it comes to the relationship between magic and science, practically everything is yet to be done.

We believe that our breakdown of the cognitive processes behind magic provides a valuable toolbox for expanding our knowledge of how cognition works. It is our hope that the work presented here will help generate a greater interest in magic as a unique research opportunity for neuroscientists, psychologists, cognitive scientists, and all who study the brain and human behavior.

Acknowledgments

The publication of this book would not have been possible without the support of the Pasqual Maragall Foundation, the Pompeu Fabra University of Barcelona, and the Institute of Neurosciences of Alicante, Spain.

The book has been enriched with valuable contributions and suggestions from friends and colleagues. For their contributions, special thanks are due to the magicians Alfredo Álvarez, Jaume Andrés, José Antonio Cachadiña, Tino Call, Dani DaOrtiz, Jesús Etcheverry, Eduard Juanola, Miguel Ángel Gea, David Sánchez, and Pipo Villanueva; and to the scientists Alex Gómez-Marín and Rodrigo Quian Quiroga. We also appreciate the editorial contribution of Hallie Stebbins and Eduardo Aparicio, the insightful comments about the history of cinema by Oscar Sánchez (Martin Pawley), the technical collaboration of Esther Román, and the comments from our partners, Rosa Cervelló and Sonia Martín.

Notes

Chapter 1. The Art and Science of the Impossible

1. Ortega Ortega, "Entrevista con 'El Monstruo.'"

2. Tamariz, "Por arte de magia."

3. Tamariz, "Teoría de la emoción mágica: Apuntes."

4. Diaconis and Graham, *Magical Mathematics.* For example, the American magician and mathematician Norman Gilbreath developed principles in the late 1950s related to the Mandelbrot set and fractal series.

5. Wagensberg, *A más cómo, menos por qué.*

Chapter 2. Living in Illusion: The Human Brain and the Visual Pathway

1. Van Leeuwen, "Book Review of *Sleights of Mind.*"

2. Otten et al., "The Uniformity Illusion."

3. Ramachandran, *A Brief Tour of Human Consciousness*; Cavanagh, "The Artist as Neuroscientist."

4. Livingstone, *Vision and Art.*

5. Ibid.

6. Ibid.

7. For those interested in Akiyoshi's work, reproductions of many of his creations can be found at http://www.ritsumei.ac.jp/~akitaoka/index-e.html.

8. Cavanagh, "The Artist as Neuroscientist."

Chapter 3. The Conception of Reality: We Are Our Memories

1. Quian Quiroga, *Borges and Memory.*

2. Von Helmholtz, "The Facts of Perception."

3. National Research Council, "Basic Research on Vision and Memory."

4. Quian Quiroga, *Qué es la memoria.*

5. Atkinson and Shiffrin, "Human Memory."

6. Baddeley, Eysenck, and Anderson, "Working Memory."

7. Engle et al., "Working Memory, Short-Term Memory, and General Fluid Intelligence."

8. Damasio, *Descartes' Error.*

9. Hoehl et al., "Itsy Bitsy Spider . . ."

Chapter 4. We Build an Illusion of Continuity

1. Livingstone, *Vision and Art.*

2. Yarbus, *Eye Movements and Vision.*

3. Anderson and Pichert, "Recall of Previously Unrecallable Information."

4. See Martinez-Conde, Macknik, and Hubel, "Microsaccadic Eye Movements"; and Rayner and Castelhano, "Eye Movements."

5. Schweitzer and Rolfs, "Intrasaccadic Motion Streaks Jump-Start Gaze Correction."

6. Intraub and Dickinson, "False Memory 1/20th of a Second Later."

7. DaOrtiz, *Reloaded.*

8. Brownlow, "Silent Films."

9. Levitin, *This is Your Brain on Music.*

10. Yao, Wood, and Simons, "As if by Magic."

11. Rensink, O'Regan, and Clark, "To See or Not to See."

12. Other magic tricks could yield different results; see Barnhart and Goldinger, "Blinded by Magic."

13. Smith, "The Role of Audience Participation."

14. Gea, *Numismagia & percepción.*

15. Thomas, Didierjean, and Nicolas, "Scientific Study of Magic."

16. Binet, "La psychologie de la prestidigitation," *Revue Philosophique de la France et de l'Étranger*; Binet, "La psychologie de la prestidigitation," *Revue des Deux Mondes.*

17. Etcheverry, *The Structural Conception of Magic.*

18. Hergovich, Gröbl, and Carbon, "The Paddle Move Commonly Used in Magic Tricks."

Chapter 5. Magic and Contrast: The Key to It All

1. Ramachandran, *The Tell-Tale Brain.*

2. Etcheverry, *The Structural Conception of Magic.*

3. Ibid.

4. Ibid.

5. Martínez et al., "Statistical Wiring of Thalamic Receptive Fields."

6. Ibid.

7. Kahneman, *Thinking, Fast and Slow.*

8. Gea, *Numismagia & percepción.*

9. DaOrtiz, *Utopia.*

10. Ortiz, *Designing Miracles.*

11. Gabi Pareras, personal communication with the author, 2018.

12. Wiseman and Greening, "'It's Still Bending'"; Wilson and French, "Magic and Memory."

13. Aronson, "Postscript."

14. Cavina-Pratesi et al., "The Magic Grasp."

15. Miguel Ángel Gea, personal communication with the author, 2018.

16. DaOrtiz, *Utopia.*

17. Baker, *Al Baker's Pet Secrets.*

18. Van de Cruys, Wagemans, and Ekroll, "The Put-and-Fetch Ambiguity."

19. Etcheverry, *The Structural Conception of Magic.*

20. Randal, "Misdirection."

21. Macknik, Martinez-Conde, and Blakeslee, *Sleights of Mind.*

22. Etcheverry, *The Structural Conception of Magic.*

Chapter 6. We Filter and Process Only What Is Useful to Us

1. Ward, "Attention."

2. Cherry, "Some Experiments on the Recognition of Speech."

3. Maskelyne and Devant, *Our Magic*; Tarbell, *The Tarbell Course in Magic*; Mulholland, *Quicker than the Eye.*

4. Tarbell, *The Tarbell Course in Magic*; Fitzkee, *Magic by Misdirection*; Bruno, *Anatomy of Misdirection*; Sharpe, *Conjurers' Psychological Secrets.*

5. Bestue, "Misdirection."

6. Leech, *Don't Look Now!*; Fitzkee, *Magic by Misdirection.*

7. Wonder and Minch, *The Books of Wonder.*

8. Ibid.

9. Lamont and Wiseman, *Magic in Theory*; Kuhn et al., "A Psychologically-Based Taxonomy of Misdirection."

10. Smith, Lamont, and Henderson, "The Penny Drops"; Smith, Lamont, and Henderson, "Change Blindness in a Dynamic Scene"; Kuhn et al., "Don't Be Fooled!"

11. Jar, *Comunicación no verbal.*

12. Kuhn et al., "Don't Be Fooled!"

13. Kuhn and Tatler, "Magic and Fixation"; Kuhn et al., "Misdirection in Magic"; Kuhn, Amlani, and Rensink, "Towards a Science of Magic."

14. Ortega et al., "Exploiting Failures in Metacognition through Magic."

15. Kuhn and Land, "There's More to Magic than Meets the Eye"; Kuhn, Tatler, and Cole, "You Look Where I Look!"; Kuhn and Rensink, "The Vanishing Ball Illusion." For Kuhn's research on attention and "misdirection," see Kuhn and Teszka, "Attention and Misdirection."

16. Lamont and Wiseman, *Magic in Theory.*

17. Thomas and Didierjean, "No Need for a Social Cue!"; Thomas and Didierjean, "The Ball Vanishes in the Air"; Tompkins, Woods, and Aimola Davies, "The Phantom Vanish Magic Trick." On audiences without this sensitivity, see Thomas and Didierjean, "Magicians Fix Your Mind."

18. Kuhn, Kourkoulou, and Leekam, "How Magic Changes Our Expectations about Autism"; Joosten et al., "Gaze and Visual Search Strategies."

19. Cui et al., "Social Misdirection Fails to Enhance a Magic Illusion"; Rieiro, Martinez-Conde, and Macknik, "Perceptual Elements in Penn & Teller's 'Cups and Balls' Magic Trick"; Tachibana and Kawabata, "The Effects of Social Misdirection"; Scott, Batten, and Kuhn, "Why Are You Looking at Me?"

20. Cui et al., "Social Misdirection Fails to Enhance a Magic Illusion."

21. Hergovich and Oberfichtner, "Magic and Misdirection"; Tamariz, *The Five Points in Magic.*

22. Cited in Macknik et al., "Attention and Awareness in Stage Magic."

23. Suchow and Alvarez, "Motion Silences Awareness of Visual Change."

24. Etcheverry, *The Structural Conception of Magic.*

25. Hergovich, Gröbl, and Carbon, "The Paddle Move Commonly Used in Magic Tricks."

26. Otero-Millán et al., "Stronger Misdirection in Curved than in Straight Motion."

27. Ibid.

28. Tachibana and Gyoba, "Effects of Different Types of Misdirection."

29. Stone, "Vision and Velocity Vectors."

30. Hyman et al., "Did You See the Unicycling Clown?"

31. Fiebelkorn, Pinsk, and Kastner, "A Dynamic Interplay within the Frontoparietal Network."

32. Simons and Chabris, "Gorillas in Our Midst."

33. Quian Quiroga, "Magic and Cognitive Neuroscience."

34. Etcheverry, *The Structural Conception of Magic.*

35. Tamariz, "Sobre la 'pregunta obnubilante.'"

36. Nobre and van Ede, "Anticipated Moments."

37. Ortiz, *Strong Magic.*

38. Barnhart et al., "Cross-Modal Attentional Entrainment."

39. Ling and Carrasco, "When Sustained Attention Impairs Perception."

40. Arciniegas, "Entrevista a Henry Evans."

41. Marcel Botella, personal communication with the author, 2015.

42. Wiseman and Nakano, "Blink and You'll Miss It."

43. Etcheverry, *The Structural Conception of Magic.*

Chapter 7. Perceiving Is a Creative Act, but Everything Is Already in Your Brain

1. Purves and Lotto, *Why We See What We Do.*

2. Intraub and Dickinson, "False Memory 1/20th of a Second Later."

3. It was Noah, not Moses, who collected animals for the Ark.

4. Evans, "How Many Pictures?"

5. Leake, "Net 'to Drain All Britain's Power.'"

6. Hoehl et al., "Itsy Bitsy Spider . . ."

7. Roskes et al., "The Right Side?"

8. Ekman, Kok, and de Lange, "Time-Compressed Preplay."

9. Whittlesea and Leboe, "The Heuristic Basis of Remembering and Classification."

10. Kuhn and Tatler, "Magic and Fixation"; Kuhn et al., "Misdirection in Magic"; Kuhn, Tatler, and Cole, "You Look Where I Look!"; Kuhn and Findlay, "Misdirection, Attention, and Awareness."

11. Cui et al., "Social Misdirection Fails to Enhance a Magic Illusion"; Otero-Millán et al., "Stronger Misdirection in Curved than in Straight Motion."

12. Wikipedia, "Gestalt Psychology," https://en.wikipedia.org/wiki/Gestalt_psychology (accessed August 8, 2020).

13. Barnhart, "The Exploitation of Gestalt Principles by Magicians."

14. Ekroll, Sayim, and Wagemans, "Against Better Knowledge"; Ekroll et al., "Illusory Visual Completion"; Ekroll and Wagemans, "Conjuring Deceptions."

15. Barnhart, "The Symmetry of Deception."

16. Ekroll et al., "Never Repeat the Same Trick Twice."

17. Van de Cruys, Wagemans, and Ekroll, "The Put-and-Fetch Ambiguity"; Ekroll and Wagemans, "Conjuring Deceptions"; Ekroll, Sayim, and Wagemans, "The Other Side of Magic."

18. Andersen et al., "The Role of Perceptual Mechanisms"; Ekroll, Sayim, and Wagemans, "The Other Side of Magic"; Øhrn et al., "A Perceptual Illusion of Empty Space."

Chapter 8. To Remember Is to Rebuild

1. Misra et al., "Minimal Memory for Details."

2. Quian Quiroga, "No Pattern Separation in the Human Hippocampus."

3. Ebbinghaus, "Memory: A Contribution to Experimental Psychology."

4. Jacobsen et al., "Why Musical Memory Can Be Preserved."

5. Sharot et al., "How Personal Experience Modulates the Neural Circuitry."

6. Hirst et al., "A Ten-Year Follow-up of a Study of Memory."

7. Redelmeier, Katz, and Kahneman, "Memories of Colonoscopy."

8. Bestue et al., "Long-Term Memory of Real-World Episodes."

9. Schacter, *The Seven Sins of Memory*.

10. Loftus and Pickrell, "The Formation of False Memories."

11. Loftus and Palmer, "Reconstruction of Automobile Destruction."

12. Roediger and McDermott, "Creating False Memories."

13. Loftus, "Eavesdropping on Memory."

14. Loftus, *Eyewitness Testimony*; Loftus, "Eavesdropping on Memory." For more on the Innocence Project, see https://www.innocenceproject.org.

15. Bestue, "Misdirection."

16. Jay, "What Do Audiences *Really* Think?"

17. Tamariz, "Fundamentos del ilusionismo."

18. Quian Quiroga, "Magic and Cognitive Neuroscience."

19. Beth and Ekroll, "The Curious Influence of Timing."

20. DaOrtiz, *Utopia*.

21. Gea, *Numismagia & percepción*.

22. Robert-Houdin, *Les secrets de la prestidigitation et de la magie*.

23. Tamariz, *The Magic Way*.

24. Ibid.

25. Prevos, *Perspectives on Magic*.

26. Ortiz, *Strong Magic*.

27. Luchins, "Mechanization in Problem Solving"; Bilalić, McLeod, and Gobet, "The Mechanism of the Einstellung (Set) Effect."

28. Danek et al., "Aha! Experiences Leave a Mark"; Danek et al., "Working Wonders?"; Danek et al., "It's a Kind of Magic"; Hedne, Norman, and Metcalfe, "Intuitive Feelings of Warmth and Confidence."

29. Tamariz, "Por arte de magia."

30. Demacheva et al., "The Applied Cognitive Psychology of Attention."

31. Ortega et al., "Exploiting Failures in Metacognition through Magic."

32. Tamariz, *The Magic Rainbow*.

33. Hodgson and Davey, "The Possibilities of Mal-observation and Lapse of Memory."

34. Besterman, "The Psychology of Testimony."

35. Houdini, *A Magician among the Spirits*.

36. Wiseman and Morris, "Recalling Pseudo-Psychic Demonstrations"; Wilson and French, "Magic and Memory."

37. Wiseman, Greening, and Smith, "Belief in the Paranormal and Suggestion"; Wiseman and Greening, "It's Still Bending'"; Wilson and French, "Magic and Memory."

38. Kahneman, *Thinking, Fast and Slow*.

39. Jay, "What Do Audiences *Really* Think?"

Chapter 9. The Undervalued Unconscious Brain

1. Eagleman, *Incognito*.

2. Cohen et al., "The Attentional Requirements of Consciousness."

3. Koch and Tsuchiya, "Attention and Consciousness"; Navajas, Nitka, and Quian Quiroga, "Dissociation between the Neural Correlates."

4. Dehaene, *Consciousness and the Brain*.

5. Gea, *Numismagia & percepción*.

6. Kawakami and Miura, "Can Magic Deception Be Detected at an Unconscious Level?"

7. Shalom et al., "Choosing in Freedom or Forced to Choose?"

8. Dehaene et al., "Imaging Unconscious Semantic Priming."

9. Dehaene, *Consciousness and the Brain*.

10. Brown, *Pure Effect*.

Chapter 10. The Magic of Decision-Making

1. For a review, see Evans, "Dual-Processing Accounts."

2. Damasio, *Descartes' Error*.

3. Kahneman, *Thinking, Fast and Slow*.

4. Ibid.

5. Melnikoff and Bargh, "The Mythical Number Two."

6. Kahneman, *Thinking, Fast and Slow*.

7. Gigerenzer, *Gut Feelings*.

8. Wan, Cheng, and Tanaka, "Neural Encoding of Opposing Strategy Values."

9. Gigerenzer, *Gut Feelings*.

10. Tversky and Kahneman, "Judgment under Uncertainty."

11. Tversky and Kahneman, "Availability."

12. Cole, "Forcing the Issue."

13. Rieznik et al., "A Massive Experiment on Choice Blindness in Political Decisions."

14. Shalom et al., "Choosing in Freedom or Forced to Choose?"; Olson et al., "Influencing Choice without Awareness."

15. Olson et al., "Simulated Thought Insertion."

16. Shalom et al., "Choosing in Freedom or Forced to Choose?"; Olson et al., "Influencing Choice without Awareness."

17. Triplett, "The Psychology of Conjuring Deceptions."

18. DaOrtiz, *Reloaded.*

19. Tino Call, personal communication with the author, 2014.

20. DaOrtiz, *Libertad de expresión.*

21. DaOrtiz, *Libertad de expresión*; DaOrtiz, *Utopia.*

22. Nisbett and Wilson, "Telling More Than We Can Know."

23. Binet, "La psychologie de la prestidigitation," *Revue Philosophique de la France et de l'Étranger*; Binet, "La psychologie de la prestidigitation," *Revue des Deux Mondes.*

24. Binet, "La psychologie de la prestidigitation," *Revue Philosophique de la France et de l'Étranger*; Binet, "La psychologie de la prestidigitation," *Revue des Deux Mondes*; Wright and Larsen, "Mental Effects with Cards"; Kubovy and Psotka, "The Predominance of Seven."

25. DaOrtiz, *Utopia.*

26. Olson, Amlani, and Rensink, "Perceptual and Cognitive Characteristics."

27. Jay, "What Do Audiences *Really* Think?"

28. McKenzie and Nelson, "What a Speaker's Choice of Frame Reveals."

29. Gigerenzer, *Gut Feelings.*

30. Moreno Bote, *¿Cómo tomamos decisiones?*

31. Johansson et al., "How Something Can Be Said about Telling More."

32. Johansson et al., "Failure to Detect Mismatches."

33. Hall et al., "Magic at the Marketplace."

34. Hall, Johansson, and Strandberg, "Lifting the Veil of Morality"; Hall et al., "How the Polls Can Be Both Spot On and Dead Wrong."

35. Johansson et al., "Choice Blindness and Preference Change."

36. Strandberg et al., "False Beliefs and Confabulation."

37. Rieznik et al., "A Massive Experiment on Choice Blindness"; Hall et al., "How the Polls Can Be Both Spot On and Dead Wrong."

38. Rieznik et al., "A Massive Experiment on Choice Blindness."

Chapter 11. The Magic Experience and Its Audiences

1. Ortiz, *Designing Miracles.*

2. Gabi Pareras, personal communication with the author, 2018.

3. Tamariz, "Teoría de la emoción mágica: Apuntes."

4. Etcheverry, *The Magic of Ascanio.*

5. Leddington, "The Experience of Magic."

6. "Penn and Teller on Broadway."

7. Ibid.

8. Nelms, *Magic and Showmanship*; Randal, "Misdirection"; Prevos, *Perspectives on Magic.*

9. Stromberg, "Teller Speaks on the Enduring Appeal of Magic."

10. Ibid.

11. Aristotle, *Poetics*; Schulz, "Fantastic Beasts and How to Rank Them."

12. Coleridge, "From *Biographia Literaria*, Chapter XIV."

13. Tolkien, "On Fairy-Stories."

14. Ortiz, *Strong Magic*.

15. Etcheverry, *The Structural Conception of Magic*.

16. Parris et al., "Imaging the Impossible."

17. Leddington, "The Experience of Magic."

18. Kuhn, *Experiencing the Impossible*.

19. Ortiz, *Strong Magic*; Ortiz, *Designing Miracles*.

20. Aronson, "Postscript."

21. Smith, Dignum, and Sonenberg, "The Construction of Impossibility."

22. Caffaratti et al., "Where Is the Ball?"

23. Hohwy, "Attention and Conscious Perception in the Hypothesis Testing Brain."

24. Grassi and Bartels, "Magic, Bayes, and Wows."

25. Leddington, "The Experience of Magic."

26. Lamont, "Magic Thinking"; Lamont, "A Particular Kind of Wonder."

27. Jay, "What Do Audiences *Really* Think?"

28. "Encuesta-estudio sobre la magia"; Jay, "What do Audiences *Really* Think?"

29. Benassi, Singer, and Reynolds, "Occult Belief"; Subbotsky, "The Belief in Magic in the Age of Science"; Kuhn, *Experiencing the Impossible*.

30. Mohr, Koutrakis, and Kuhn, "Priming Psychic and Conjuring Abilities"; Lesaffre et al., "Magic Performances."

31. Lan et al., "Fake Science."

32. Parris et al., "Imaging the Impossible"; Danek et al., "An fMRI Investigation of Expectation Violation."

33. Danek et al., "An fMRI Investigation of Expectation Violation."

34. Williams and McOwan, "Magic in the Machine"; Williams and McOwan, "Magic in Pieces"; Williams and McOwan, "The Magic Words."

35. Jay, "What Do Audiences *Really* Think?"

36. Ibid.

37. Griffiths, "Revealing Ontological Commitments by Magic."

38. Zaghi-Lara et al., "Playing Magic Tricks."

39. Gea, *Numismagia and Percepción*.

40. Mulholland, *Quicker than the Eye*.

41. Binet, "La psychologie de la prestidigitation," *Revue Philosophique de la France et de l'Étranger*; Binet, "La psychologie de la prestidigitation," *Revue des Deux Mondes*.

42. Hay et al., "Using Drawings of the Brain Cell to Exhibit Expertise in Neuroscience."

43. Kuhn and Teszka, "Don't Get Misdirected!"

44. Lewry et al., "Intuitions about Magic Track the Development of Intuitive Physics."

45. Olson, Demacheva, and Raz, "Explanations of a Magic Trick across the Life Span."

46. Aragón, *Dr. Woody*.

47. Mago Koko, personal communication with the author, 2017.

48. Silvia et al., "Who Hates Magic?"

49. Stromberg, "Teller Speaks on the Enduring Appeal of Magic."

50. Seneca, *Seneca's Letters from a Stoic*, 102.

51. Kuhn and Findlay, "Misdirection, Attention, and Awareness."

52. Houdini, *A Magician among the Spirits.*

53. Nardi, "The Social World of Magicians."

54. Steinkraus, "The Art of Conjuring."

Chapter 12. Scientific Research and Magic

1. Robert-Houdin, *Les secrets de la prestidigitation et de la magie*; Dessoir, "The Psychology of Legerdemain"; Maskelyne and Devant, *Our Magic.*

2. Jay, *Magic in Mind.*

3. Robert-Houdin, *Les secrets de la prestidigitation et de la magie.*

4. Richard, *El mago de los salones o el diablo color de rosa.*

5. Lachapelle, "From the Stage to the Laboratory"; Lamont, "The Misdirected Quest."

6. Binet, "La psychologie de la prestidigitation," *Revue Philosophique de la France et de l'Étranger*; Binet, "La psychologie de la prestidigitation," *Revue des Deux Mondes.*

7. Dessoir, "The Psychology of Legerdemain."

8. Jastrow, "The Psychology of Deception"; Triplett, "The Psychology of Conjuring Deceptions."

9. Jastrow, "Psychological Notes upon Sleight-of-Hand Experts."

10. Triplett, "The Psychology of Conjuring Deceptions."

11. De Vere, *Petite magie blanche.*

12. Richard, *El mago de los salones o el diablo color de rosa.*

13. Von Helmholtz, "The Facts of Perception"; and Gregory, "Conjuring."

14. Kuhn and Tatler, "Magic and Fixation."

15. Kuhn, Amlani, and Rensink, "Towards a Science of Magic."

16. Kuhn and Martínez, "Misdirection: Past, Present, and the Future."

17. Kuhn et al., "A Psychologically-Based Taxonomy of Misdirection."

18. Rensink and Kuhn, "A Framework for Using Magic to Study the Mind"; Rensink and Kuhn, "The Possibility of a Science of Magic"; Kuhn, *Experiencing the Impossible.*

19. Thomas et al., "Does Magic Offer a Cryptozoology Ground for Psychology?"

20. Macknik et al., "Attention and Awareness in Stage Magic."

21. Macknik, Martinez-Conde, and Blakeslee, *Sleights of Mind.*

22. Lamont and Henderson, "More Attention and Greater Awareness in the Scientific Study of Magic"; Macknik and Martinez-Conde, "Real Magic."

23. Lamont, Henderson, and Smith, "Where Science and Magic Meet"; Lamont, "Problems with the Mapping of Magic Tricks."

24. Lamont, Henderson, and Smith, "Where Science and Magic Meet."

25. Camí, Gómez-Marín, and Martínez, "On the Cognitive Bases of Illusionism."

26. Tompkins, "A Science of Magic Bibliography—2021 Update."

27. Raz, Olson, and Kuhn, "The Psychology of Magic and the Magic of Psychology."

28. Science of Magic Association, "Past Events," https://scienceofmagicassoc.org/home/ (accessed September 22, 2020).

29. Ibid.

30. Gómez-Marín, Martínez, and Camí, "Science as Magic."

31. Tompkins, "A Science of Magic Bibliography—2021 Update."

Bibliography

Andersen, Steffen, Anna Ring, Mats Svalebjørg, and Vebjørn Ekroll. "The Role of Perceptual Mechanisms in Creating the Illusion of Levitation." Paper presented at the Science of Magic Association meeting, Goldsmiths University of London, August 31–September 1, 2017.

Anderson, Richard C., and James W. Pichert. "Recall of Previously Unrecallable Information Following a Shift in Perspective." *Journal of Verbal Learning and Verbal Behavior* 17, no. 1 (February 1978): 1–12.

Aragón, Woody. *Dr. Woody: Los terribles experimentos cartomágicos del malvado Woody Aragón.* Madrid: Self-published, 2018.

Arciniegas, Germán. "Entrevista a Henry Evans, las enseñanzas de un campeón del mundo." May 3, 2017. https://pastomagic.com/2017/05/entrevista-a-henry-evans-las-ensenanzas-de-un-campeon-del-mundo/.

Aristotle. *Poetics*, translated by Malcolm Heath. New York: Penguin Books, 1996.

Aronson, Simon. "Postscript." In Aronson, *The Aronson Approach*, 172. Chicago: Savaco Ltd., 1990.

Atkinson, Richard C., and Richard M. Shiffrin. "Human Memory: A Proposed System and Its Control Processes." In *The Psychology of Learning and Motivation: Advances in Research and Theory*, vol. 2, edited by Kenneth W. Spence and Janet T. Spence, 89–195. New York: Academic Press, 1968.

Baddeley, Alan, Michael W. Eysenck, and Michael C. Anderson. "Working Memory." In Baddeley, Eysenck, and Anderson, *Memory*, 2nd ed., 67–105. New York: Psychology Press, 2015.

Baker, Al. *Al Baker's Pet Secrets*. Minneapolis: Carl W. Jones, 1951.

Barnhart, Anthony S. "The Exploitation of Gestalt Principles by Magicians." *Perception* 39, no. 9 (2010): 1286–89.

———. "The Symmetry of Deception: Predictability Reduces Attention toward Symmetrical Actions." Paper presented at the Science of Magic Association meeting, Goldsmiths University of London, August 31–September 1, 2017.

Barnhart, Anthony S., Mandy J. Ehlert, Stephen D. Goldinger, and Alison D. Mackey. "Cross-Modal Attentional Entrainment: Insights from Magicians." *Attention, Perception, and Psychophysics* 80, no. 5 (2018): 1240–49.

Barnhart, Anthony S., and Stephen D. Goldinger. "Blinded by Magic: Eye-Movements Reveal the Misdirection of Attention." *Frontiers in Psychology* 5, no. 1461 (December 2014): 1–9.

Benassi, Victor A., Barry Singer, and Craig B. Reynolds. "Occult Belief: Seeing Is Believing." *Journal for the Scientific Study of Religion* 19, no. 4 (1980): 337–49.

Besterman, Theodore. “The Psychology of Testimony in Relation to Paraphysical Phenomena: Report of an Experiment.” *Proceedings of the Society for Psychical Research* 40 (1932): 363–87.

Bestue, David. “Misdirection: Model of Information Process for Neuromagic.” *PsyArXiv Preprints*, April 23, 2020. doi:10.31234/osf.io/vg6y9.

Bestue, David, Luis M. Martínez, Alex Gómez-Marín, Miguel Ángel Gea, and Jordi Camí. “Long-Term Memory of Real-World Episodes Is Independent of Recency Effects: Magic Tricks as Ecological Tasks.” *Heliyon* 6, no. 10 (2020): e05260.

Beth, Tessa, and Vebjørn Ekroll. “The Curious Influence of Timing on the Magical Experience Evoked by Conjuring Tricks Involving False Transfer: Decay of Amodal Object Permanence?” *Psychological Research* 79, no. 4 (2015): 513–22.

Bilalić, Merim, Peter McLeod, and Fernand Gobet. “The Mechanism of the Einstellung (Set) Effect: A Pervasive Source of Cognitive Bias.” *Current Directions in Psychological Science* 19, no. 2 (2010): 111–15.

Binet, Alfred. “La psychologie de la prestidigitation.” *Revue Philosophique de la France et de l’Étranger* 37 (March 1894): 346–48.

———. “La psychologie de la prestidigitation.” *Revue des Deux Mondes* 125 (1894): 903–22.

Brossa, Joan. *Vivarium*. Barcelona: Edicions 62, 1972.

Brown, Derren. *Pure Effect: Direct Mindreading and Magical Artistry*. Humble, TX: H&R Magic Books, 2000.

Brownlow, Kevin. “Silent Films. What Was the Right Speed?” *Sight and Sound* (Summer 1980): 164–67. https://web.archive.org/web/20110724032550/http://www.cinemaweb.com/silentfilm/bookshelf/18_kb_2.htm.

Bruno, Joe. *Anatomy of Misdirection*. Baltimore: Stoney Brook Press, 1978.

Caffaratti, Hugo, Joaquin Navajas, Hernán G. Rey, and Rodrigo Quian Quiroga. “Where Is the Ball? Behavioral and Neural Responses Elicited by a Magic Trick.” *Psychophysiology* 53, no. 9 (2016): 1441–48.

Camí, Jordi, Alex Gómez-Marín, and Luis M. Martínez. “On the Cognitive Bases of Illusionism.” *PeerJ* 8 (2020): e9712.

Cavanagh, Patrick. “The Artist as Neuroscientist.” *Nature* 434 (March 2005): 301–7.

Cavina-Pratesi, Cristiana, Gustav Kuhn, Magdalena Ietswaart, and A. David Milner. “The Magic Grasp: Motor Expertise in Deception.” *PLOS ONE* 6, no. 2 (February 2011): 1–5.

Cherry, E. Colin. “Some Experiments on the Recognition of Speech, with One and with Two Ears.” *Journal of the Acoustical Society of America* 25, no. 5 (1953): 975–79.

chrisj01. “Hitchcock’s Elevator Story.” October 7, 2011. https://www.youtube.com/watch?v=PqXFtWSBBd4&app=desktop.

Cohen, Michael A., Patrick Cavanagh, Marvin M. Chun, and Ken Nakayama. “The Attentional Requirements of Consciousness.” *Trends in Cognitive Sciences* 16, no. 8 (2012): 411–17.

Cole, Geoff G. “Forcing the Issue: Little Psychological Influence in a Magician’s Paradigm.” *Consciousness and Cognition* 84 (2020): 103002.

Coleridge, Samuel Taylor. “From *Biographia Literaria*, Chapter XIV.” Poetry Foundation, October 13, 2009. https://www.poetryfoundation.org/articles/69385/from-biographia-literaria-chapter-xiv.

Cui, Jie, Jorge Otero-Millan, Stephen L. Macknik, Mac King, and Susana Martinez-Conde. "Social Misdirection Fails to Enhance a Magic Illusion." *Frontiers in Human Neuroscience* 5, no. 103 (2011): 1–11.

Damasio, Antonio R. *Descartes' Error: Emotion, Reason, and the Human Brain*. New York: G. P. Putnam, 1994.

Danek, Amory H., Thomas Fraps, Albrecht von Müller, Benedikt Grothe, and Michael Öllinger. "Aha! Experiences Leave a Mark: Facilitated Recall of Insight Solutions." *Psychological Research* 77 (2013): 659–69.

———. "Working Wonders? Investigating Insight with Magic Tricks." *Cognition* 130, no. 2 (2014): 174–85.

———. "It's a Kind of Magic—What Self-Reports Can Reveal about the Phenomenology of Insight Problem Solving." *Frontiers in Psychology* 5, no. 1408 (2014): 1–11.

Danek, Amory H., Michael Öllinger, Thomas Fraps, Benedikt Grothe, and Virginia L. Flanagin. "An fMRI Investigation of Expectation Violation in Magic Tricks." *Frontiers in Psychology* 6, no. 84 (2015): 1–11.

DaOrtiz, Dani. *Libertad de expresión: El forzaje psicológico y la psicología en el forzaje*. Málaga: GrupoKaps, 2009.

———. *Utopia*. Essential Magic Collection (4 DVDs), edited by Luis de Matos. Ansião, Portugal: Luis de Matos Produções. 2011.

———. *Reloaded*. Essential Magic Collection (4 DVDs), edited by Luis de Matos. Ansião, Portugal: Luis de Matos Produções, 2015.

Dehaene, Stanislas. *Consciousness and the Brain: Deciphering How the Brain Codes Our Thoughts*. New York: Penguin Books, 2014.

Dehaene, Stanislas, Lionel Naccache, Gurvan Le Clec'H, Etienne Koechlin, Michael Mueller, Ghislaine Dehaene-Lambertz, Pierre-François van de Moortele, and Denis Le Bihan. "Imaging Unconscious Semantic Priming." *Nature* 395 (1998): 597–600.

Demacheva, Irina, Martin Ladouceur, Ellis Steinberg, Galina Pogossova, and Amir Raz. "The Applied Cognitive Psychology of Attention: A Step Closer to Understanding Magic Tricks." *Applied Cognitive Psychology* 26, no. 4 (2012): 541–49.

Dessoir, Max. "The Psychology of Legerdemain." *The Open Court* 12 (1893): 599–606.

De Vere, Charles. *Petite magie blanche ou tours de physique amusante dévoilés*. Paris: Delarue, 1879.

Diaconis, Persi, and Ron Graham. *Magical Mathematics: The Mathematical Ideas That Animate Great Magic Tricks*. Princeton, NJ: Princeton University Press, 2012.

Eagleman, David. *Incognito: The Secret Lives of the Brain*. New York: Pantheon Books, 2011.

Ebbinghaus, Hermann. "Memory: A Contribution to Experimental Psychology," translated by Henry A. Ruger and Clara E. Bussenius. *Annals of Neurosciences* 20, no. 4 (2013): 155–56. Originally published as *Über das Gedächtnis: Untersuchungen zur experimentellen Psychologie* (Leipzig: Duncker & Humblot, 1885).

Edwards, Adrian, Glyn Elwyn, Judith Covey, Elaine Matthews, and Rolsin Pill. "Presenting Risk Information: A Review of the Effects of 'Framing' and Other Manipulations on Patient Outcomes." *Journal of Health Communication* 6, no. 1 (2001): 61–82.

Ekman, Matthias, Peter Kok, and Floris P. de Lange. "Time-Compressed Preplay of Anticipated Events in Human Primary Visual Cortex." *Nature Communications* 8, no. 15276 (2017): 1–9.

Ekroll, Vebjørn, Evy De Bruyckere, Lotte Vanwezemael, and Johan Wagemans. "Never Repeat the Same Trick Twice—Unless It Is Cognitively Impenetrable." *i-Perception* 9, no. 6 (2018): 1–14.

Ekroll, Vebjørn, Bilge Sayim, Ruth Van der Hallen, and Johan Wagemans. "Illusory Visual Completion of an Object's Invisible Backside Can Make Your Finger Feel Shorter." *Current Biology* 26, no. 8 (2016): 1029–33.

Ekroll, Vebjørn, Bilge Sayim, and Johan Wagemans. "Against Better Knowledge: The Magical Force of Amodal Volume Completion." *i-Perception* 4, no. 8 (2013): 511–15.

———. "The Other Side of Magic: The Psychology of Perceiving Hidden Things." *Perspectives on Psychological Science* 12, no. 1 (2017): 91–106.

Ekroll, Vebjørn, and Johan Wagemans. "Conjuring Deceptions: Fooling the Eye or Fooling the Mind?" *Trends in Cognitive Sciences* 20, no. 7 (July 2016): 486–89.

"Encuesta-estudio sobre la magia." *Circular de la Escuela Mágica de Madrid* 1 (April 1975).

Engle, Randall W., Stephen W. Tuholski, James E. Laughlin, and Andrew R. A. Conway. "Working Memory, Short-Term Memory, and General Fluid Intelligence: A Latent-Variable Approach." *Journal of Experimental Psychology: General* 128, no. 3 (1999): 309–31.

Etcheverry, Jesús. *The Structural Conception of Magic*, vol. 1 of *The Magic of Ascanio*. Madrid: Páginas, 2005.

Evans, Benedict. "How Many Pictures?" August 27, 2015. https://www.ben-evans.com/benedictevans/2015/8/19/how-many-pictures.

Evans, Jonathan St. B. T. "Dual-Processing Accounts of Reasoning, Judgment, and Social Cognition." *Annual Review of Psychology* 59 (2008): 255–78.

Fiebelkorn, Ian C., Mark A. Pinsk, and Sabine Kastner. "A Dynamic Interplay within the Frontoparietal Network Underlies Rhythmic Spatial Attention." *Neuron* 99, no. 4 (2018): 842–53.

Fitzkee, Dariel. *Magic by Misdirection* (1945). Pomeroy, OH: Lee Jacobs Productions, 1987.

Fraps, Thomas. "Time and Magic—Manipulating Subjective Temporality." In *Subjective Time: The Philosophy, Psychology, and Neuroscience of Temporality*, edited by Valtteri Arstila and Dan Lloyd, 263–86. Cambridge, MA: MIT Press, 2014.

Gea, Miguel Ángel. *Numismagia & percepción*. Madrid: Taranco Producciones, 2018.

Gigerenzer, Gerd. *Gut Feelings: The Intelligence of the Unconscious*. New York: Penguin Books, 2007.

Gómez-Marín, Alex, Luis M. Martínez, and Jordi Camí. "Science as Magic." *Organisms* 4, no. 1 (2020): 90–101.

Grassi, Pablo R., and Andreas Bartels. "Magic, Bayes, and Wows: A Bayesian Account of Magic Tricks." *Neuroscience and Biobehavioral Reviews* 126 (2021): 515–27.

Gregory, Richard L. "Conjuring." *Perception* 11 (1982): 631–33.

Griffiths, Thomas L. "Revealing Ontological Commitments by Magic." *Cognition* 136 (2015): 43–48.

Hall, Lars, Petter Johansson, and Thomas Strandberg. "Lifting the Veil of Morality: Choice Blindness and Attitude Reversals on a Self-Transforming Survey." *PLOS ONE* 7, no. 9 (2012): e45457.

Hall, Lars, Petter Johansson, Betty Tärning, Sverker Sikström, and Thérèse Deutgen. "Magic at the Marketplace: Choice Blindness for the Taste of Jam and the Smell of Tea." *Cognition* 117, no. 1 (2010): 54–61.

Hall, Lars, Thomas Strandberg, Philip Pärnamets, Andreas Lind, Betty Tärning, and Petter Johansson. "How the Polls Can Be Both Spot On and Dead Wrong: Using Choice Blindness to Shift Political Attitudes and Voter Intentions." *PLOS ONE* 8, no. 4 (2013): e60554.

Hay, David B., Darren Williams, Daniel Stahl, and Richard J. Wingate. "Using Drawings of the Brain Cell to Exhibit Expertise in Neuroscience: Exploring the Boundaries of Experimental Culture." *Science Education* 97, no. 3 (2013): 468–91.

Hebb, Donald. *The Organization of Behavior: A Neuropsychological Theory*. New York: Wiley & Sons, 1949.

Hedne, Mikael R., Elisabeth Norman, and Janet Metcalfe. "Intuitive Feelings of Warmth and Confidence in Insight and Noninsight Problem Solving of Magic Tricks." *Frontiers in Psychology* 7, no. 1314 (2016): 1–13.

Hergovich, Andreas, Kristian Gröbl, and Claus-Christian Carbon. "The Paddle Move Commonly Used in Magic Tricks as a Means for Analysing the Perceptual Limits of Combined Motion Trajectories." *Perception* 40, no. 3 (January 2011): 358–66.

Hergovich, Andreas, and Bernhard Oberfichtner. "Magic and Misdirection: The Influence of Social Cues on the Allocation of Visual Attention While Watching a Cups-and-Balls Routine." *Frontiers in Psychology* 7, no. 761 (2016): 1–16.

Hirst, William, Elizabeth A. Phelps, Robert Meksin, Chandan J. Vaidya, Marcia K. Johnson, Karen J. Mitchell, Randy L. Buckner, Andrew E. Budson, John D. E. Gabrieli, Cindy Lustig, et al. "A Ten-Year Follow-Up of a Study of Memory for the Attack of September 11, 2001: Flashbulb Memories and Memories for Flashbulb Events." *Journal of Experimental Psychology: General* 144, no. 3 (2015): 604–23.

Hodgson, Richard, and Samuel John Davey. "The Possibilities of Mal-observation and Lapse of Memory from a Practical Point of View." *Proceedings of the Society for Psychical Research* 4 (1887): 381–495.

Hoehl, Stefanie, Kahl Hellmer, Maria Johansson, and Gustaf Gredebäck. "Itsy Bitsy Spider . . . : Infants React with Increased Arousal to Spiders and Snakes." *Frontiers in Psychology* 8, no. 1710 (October 2017): 1–8.

Hohwy, Jakob. "Attention and Conscious Perception in the Hypothesis Testing Brain." *Frontiers in Psychology* 3, no. 96 (2012): 1–14.

Houdini, Harry. *A Magician among the Spirits*. New York: Harper & Brothers, 1924.

Hyman, Ira E., Jr., S. Matthew Boss, Breanne M. Wise, Kira E. McKenzie, and Jenna M. Caggiano. "Did You See the Unicycling Clown? Inattentional Blindness While Walking and Talking on a Cell Phone." *Applied Cognitive Psychology* 24, no. 5 (2010): 597–607.

Intraub, Helene, and Christopher A. Dickinson. "False Memory 1/20th of a Second Later: What the Early Onset of Boundary Extension Reveals about Perception." *Psychological Science* 19, no. 10 (October 2008): 1007–14.

Jacobsen, Jörn-Henrik, Johannes Stelzer, Thomas Hans Fritz, Gael Chételat, Renaud La Joie, and Robert Turner. "Why Musical Memory Can Be Preserved in Advanced Alzheimer's Disease." *Brain* 138, no. 8 (2015): 2438–50.

Jar, Núria. *Comunicación no verbal: El lenguaje silencioso, más allá de las palabras*. Barcelona: EMSE EDAPP, 2018.

Jastrow, Joseph. "Psychological Notes upon Sleight-of-Hand Experts." *Science* 3, no. 71 (1896): 685–89.

Jastrow, Joseph. "The Psychology of Deception." *Popular Science Monthly* 34 (December 1888): 145–57.

———. "Studies from the Laboratory of Experimental Psychology of the University of Wisconsin—II." *American Journal of Psychology* 4, no. 3 (1892): 381–428.

———. "Psychological Notes upon Sleight-of-Hand Experts." *Science* 3, no. 71 (1896): 685–89.

Jay, Joshua, ed. *Magic in Mind: Essential Essays for Magicians*. Gloucester, UK: Vanishing, 2013.

———. "What Do Audiences *Really* Think?" *Magic* 25, no. 13 (September 2016): 46–55.

Johansson, Petter, Lars Hall, Sverker Sikström, and Andreas Olsson. "Failure to Detect Mismatches between Intention and Outcome in a Simple Decision Task." *Science* 310, no. 5745 (2005): 116–19.

Johansson, Petter, Lars Hall, Sverker Sikström, Betty Tärning, and Andreas Lind. "How Something Can Be Said about Telling More than We Can Know: On Choice Blindness and Introspection." *Consciousness and Cognition* 15, no. 4 (2006): 673–92.

Johansson, Petter, Lars Hall, Betty Tärning, Sverker Sikström, and Nick Chater. "Choice Blindness and Preference Change: You Will Like This Paper Better if You (Believe You) Chose to Read It!" *Journal of Behavioral Decision Making* 27, no. 3 (2014): 281–89.

Joosten, Annette, Sonya Girdler, Matthew A. Albrecht, Chiara Horlin, Marita Falkmer, Denise Leung, Anna Ordqvist, Håkan Fleischer, and Torbjörn Falkmer. "Gaze and Visual Search Strategies of Children with Asperger Syndrome/High Functioning Autism Viewing a Magic Trick." *Developmental Neurorehabilitation* 19, no. 2 (2016): 95–102.

Kahneman, Daniel. *Thinking, Fast and Slow*. New York: Farrar, Straus and Giroux, 2011.

Kawakami, Naoaki, and Emi Miura. "Can Magic Deception Be Detected at an Unconscious Level?" *Perception* 46, no. 6 (2017): 698–708.

Koch, Christof, and Naotsugu Tsuchiya. "Attention and Consciousness: Two Distinct Brain Processes." *Trends in Cognitive Sciences* 11, no. 1 (2007): 16–22.

Kubovy, Michael, and Joseph Psotka. "The Predominance of Seven and the Apparent Spontaneity of Numerical Choices." *Journal of Experimental Psychology: Human Perception and Performance* 2, no. 2 (1976): 291–94.

Kuhn, Gustav. *Experiencing the Impossible: The Science of Magic*. Cambridge, MA: MIT Press, 2019.

Kuhn, Gustav, Alym A. Amlani, and Ronald A. Rensink. "Towards a Science of Magic." *Trends in Cognitive Sciences* 12, no. 9 (2008): 349–54.

Kuhn, Gustav, Hugo A. Caffaratti, Robert Teszka, and Ronald A. Rensink. "A Psychologically-Based Taxonomy of Misdirection." *Frontiers in Psychology* 5, no. 1392 (2014): 1–14.

Kuhn, Gustav, and John M. Findlay. "Misdirection, Attention, and Awareness: Inattentional Blindness Reveals Temporal Relationship between Eye Movements and Visual Awareness." *Quarterly Journal of Experimental Psychology* 63, no. 1 (2010): 136–46.

Kuhn, Gustav, Anastasia Kourkoulou, and Susan R. Leekam. "How Magic Changes Our Expectations about Autism." *Psychological Science* 21, no. 10 (2010): 1487–93.

Kuhn, Gustav, and Michael F. Land. "There's More to Magic than Meets the Eye." *Current Biology* 16, no. 22 (2006): R950–51.

Kuhn, Gustav, and Luis M. Martínez. "Misdirection: Past, Present, and the Future." *Frontiers in Human Neuroscience* 5, no. 172 (2012): 1–6.

Kuhn, Gustav, and Ronald A. Rensink. "The Vanishing Ball Illusion: A New Perspective on the Perception of Dynamic Events." *Cognition* 148 (March 2016): 64–70.

Kuhn, Gustav, and Benjamin W. Tatler. "Magic and Fixation: Now You Don't See It, Now You Do." *Perception* 34, no. 9 (2005): 1155–61.

Kuhn, Gustav, Benjamin W. Tatler, and Geoff G. Cole. "You Look Where I Look! Effect of Gaze Cues on Overt and Covert Attention in Misdirection." *Visual Cognition* 17, nos. 6–7 (2009): 925–44.

Kuhn, Gustav, Benjamin W. Tatler, John M. Findlay, and Geoff G. Cole. "Misdirection in Magic: Implications for the Relationship between Eye Gaze and Attention." *Visual Cognition* 16, nos. 2–3 (2008): 391–405.

Kuhn, Gustav, and Robert Teszka. "Attention and Misdirection: How to Use Conjuring Experience to Study Attentional Processes." In *The Handbook of Attention*, edited by Jonathan M. Fawcett, Evan F. Risko, and Alan Kingstone, 503–23. Cambridge, MA: MIT Press, 2015.

———. "Don't Get Misdirected! Differences in Overt and Covert Attentional Inhibition between Children and Adults." *Quarterly Journal of Experimental Psychology* 71, no. 3 (2018): 688–94.

Kuhn, Gustav, Robert Teszka, Natalia Tenaw, and Alan Kingstone. "Don't Be Fooled! Attentional Responses to Social Cues in a Face-to-Face and Video Magic Trick Reveals Greater Top-Down Control for Overt than Covert Attention." *Cognition* 146 (January 2016): 136–42.

Lachapelle, Sofie. "From the Stage to the Laboratory: Magicians, Psychologists, and the Science of Illusion." *Journal of the History of the Behavioral Sciences* 44, no. 4 (2008): 319–34.

Lamont, Peter. "The Misdirected Quest." *The Psychologist* 23, no. 12 (2010): 978–80.

———. "Problems with the Mapping of Magic Tricks." *Frontiers in Psychology* 6, no. 855 (2015): 1–3.

———. "A Particular Kind of Wonder: The Experience of Magic Past and Present." *Review of General Psychology* 21, no. 1 (2017): 1–8.

———. "Magical Thinking." *Genii* 80, no. 7 (2017): 74–79.

Lamont, Peter, and John M. Henderson. "More Attention and Greater Awareness in the Scientific Study of Magic." *Nature Reviews Neuroscience* 10 (2009): 241. https://doi.org/10.1038/nrn2473-c1.

Lamont, Peter, John M. Henderson, and Tim J. Smith. "Where Science and Magic Meet: The Illusion of a 'Science of Magic.'" *Review of General Psychology* 14, no. 1 (2010): 16–21.

Lamont, Peter, and Richard Wiseman. *Magic in Theory*. Seattle: Hermetic Press, 1999.

Lan, Yuxuan, Christine Mohr, Xiaomeng Hu, and Gustav Kuhn. "Fake Science: The Impact of Pseudo-Psychological Demonstrations on People's Beliefs in Psychological Principles." *PLOS ONE* 13, no. 11 (2018): e0207629.

Leake, Jonathan. "Net 'to Drain All Britain's Power.'" *Sunday Times*, May 3, 2015. https://www.thetimes.co.uk/article/net-to-drain-all-britains-power-prm2qx8czp0.

Leddington, Jason. "The Experience of Magic." *Journal of Aesthetics and Art Criticism* 74, no. 3 (201): 253–64.

Leech, Al. *Don't Look Now! The Smart Slant on Misdirection*. Chicago: Ireland Magic Co., 1960.

Lesaffre, Lise, Gustav Kuhn, Ahmad Abu-Akel, Déborah Rochat, and Christine Mohr. "Magic Performances—When Explained in Psychic Terms by University Students." *Frontiers in Psychology* 9, no. 2129 (2018): 1–12.

Levitin, Daniel J. *This Is Your Brain on Music: The Science of a Human Obsession.* Dutton: Penguin Books, 2006.

Lewry, Casey, Kaley Curtis, Nadya Vasilyeva, Fei Xu, and Thomas L. Griffiths. "Intuitions about Magic Track the Development of Intuitive Physics." *Cognition* 214 (2021): 104762.

Libet, Benjamin. "Unconscious Cerebral Initiative and the Role of Conscious Will in Voluntary Action." *Behavioral and Brain Sciences* 8, no. 4 (1985): 529–66.

Ling, Sam, and Marisa Carrasco. "When Sustained Attention Impairs Perception." *Nature Neuroscience* 9, no. 10 (2006): 1243–45.

Livingstone, Margaret, *Vision and Art: The Biology of Seeing.* New York: Harry N. Abrams, 2002.

Loftus, Elizabeth F. *Eyewitness Testimony.* Cambridge, MA: Harvard University Press, 1979.

———. "Eavesdropping on Memory." *Annual Review of Psychology* 68 (2017): 1–18.

Loftus, Elizabeth F., and John C. Palmer. "Reconstruction of Automobile Destruction: An Example of the Interaction between Language and Memory." *Journal of Verbal Learning and Verbal Behavior* 13, no. 5 (1974): 585–89.

Loftus, Elizabeth F., and Jacqueline E. Pickrell. "The Formation of False Memories." *Psychiatric Annals* 25 (1995): 720–25.

Luchins, Abraham S. "Mechanization in Problem Solving: The Effect of Einstellung." *Psychological Monographs* 54, no. 6 (1942): i–95.

Macknik, Stephen L., Mac King, James Randi, Apollo Robbins, Teller, John Thompson, and Susana Martinez-Conde. "Attention and Awareness in Stage Magic: Turning Tricks into Research." *Nature Reviews Neuroscience* 9 (2008): 871–79. https://doi.org/10.1038/nrn2473.

Macknik, Stephen L., and Susana Martinez-Conde. "Real Magic: Future Studies of Magic Should Be Grounded in Neuroscience." *Nature Reviews Neuroscience* 10 (2009): 241. https://doi.org/10.1038/nrn2473-c2.

Macknik, Stephen L., Susana Martinez-Conde, and Sandra Blakeslee. *Sleights of Mind: What the Neuroscience of Magic Reveals about Our Everyday Deceptions.* New York: Henry Holt and Co., 2010.

Márquez, Gabriel García. *Love in the Time of Cholera.* New York: Alfred A. Knopf, 1988.

Martínez, Luis M., Manuel Molano-Mazón, Xin Wang, Friedrich T. Sommer, and Judith A. Hirsch. "Statistical Wiring of Thalamic Receptive Fields Optimizes Spatial Sampling of the Retinal Image." *Neuron* 81, no. 4 (February 2014): 943–56.

Martinez-Conde, Susana, Stephen L. Macknik, and David H. Hubel. "Microsaccadic Eye Movements and Firing of Single Cells in the Striate Cortex of Macaque Monkeys." *Nature Neuroscience* 3, no. 3 (March 2000): 251–58.

Maskelyne, Nevil, and David Devant. *Our Magic.* New York: E. P. Dutton & Co., 1911.

McKenzie, Craig R. M., and Jonathan D. Nelson. "What a Speaker's Choice of Frame Reveals: Reference Points, Frame Selection, and Framing Effects." *Psychonomic Bulletin and Review* 10, no. 3 (2003): 596–602.

Melnikoff, David E., and John A. Bargh. "The Mythical Number Two." *Trends in Cognitive Sciences* 22, no. 4 (2018): 280–93.

Misra, Pranav, Alyssa Marconi, Matthew Peterson, and Gabriel Kreiman. "Minimal Memory for Details in Real Life Events." *Scientific Reports* 8, no. 16701 (2018). https://doi.org/10.1038/s41598-018-33792-2.

Mohr, Christine, Nikolaos Koutrakis, and Gustav Kuhn. "Priming Psychic and Conjuring Abilities of a Magic Demonstration Influences Event Interpretation and Random Number Generation Biases." *Frontiers in Psychology* 5, no. 1542 (2015): 1–8.

Moreno Bote, Rubén. *¿Cómo tomamos decisiones? Los mecanismos neuronales de la elección.* Barcelona: EMSE EDAPP SL, 2018.

Mulholland, John. *Quicker than the Eye: The Magic and Magicians of the World.* Indianapolis: Bobbs-Merrill Co., 1932.

Nardi, Peter M. "The Social World of Magicians: Gender and Conjuring." *Sex Roles: A Journal of Research* 19, nos. 11–12 (1988): 759–70.

National Research Council. "Basic Research on Vision and Memory." In *Identifying the Culprit: Assessing Eyewitness Identification*, 45–70. Washington, DC: National Academies Press, 2014.

Navajas, Joaquín, Aleksander W. Nitka, and Rodrigo Quian Quiroga. "Dissociation between the Neural Correlates of Conscious Face Perception and Visual Attention." *Psychophysiology* 54, no. 8 (2017): 1138–50.

Nelms, Henning. *Magic and Showmanship: A Handbook for Conjurers.* New York: Dover Publications, 1969.

Nisbett, Richard E., and Timothy D. Wilson. "Telling More than We Can Know: Verbal Reports on Mental Processes." *Psychological Review* 84, no. 3 (1977): 231–59.

Nobre, Anna C., and Freek van Ede. "Anticipated Moments: Temporal Structure in Attention." *Nature Reviews Neuroscience* 19 (2018): 34–48. https://doi.org/10.1038/nrn.2017.141.

Øhrn, Heidi, Mats Svalebjørg, Steffen Andersen, Anna Edit Ring, and Vebjørn Ekroll. "A Perceptual Illusion of Empty Space Can Create a Perceptual Illusion of Levitation." *i-Perception* 10, no. 6 (2019): 1–16.

Olson, Jay A., Alym A. Amlani, Amir Raz, and Ronald A. Rensink. "Influencing Choice without Awareness." *Consciousness and Cognition* 37 (2015): 225–36.

Olson, Jay A., Alym A. Amlani, and Ronald A. Rensink. "Perceptual and Cognitive Characteristics of Common Playing Cards." *Perception* 41, no. 3 (2012): 268–86.

Olson, Jay A., Irina Demacheva, and Amir Raz. "Explanations of a Magic Trick across the Life Span." *Frontiers in Psychology* 6, no. 219 (2015): 1–5.

Olson, Jay A., Mathieu Landry, Krystèle Appourchaux, and Amir Raz. "Simulated Thought Insertion: Influencing the Sense of Agency Using Deception and Magic." *Consciousness and Cognition* 43 (2016): 11–26.

Ortega, Jeniffer, Patricia Montañes, Anthony Barnhart, and Gustav Kuhn. "Exploiting Failures in Metacognition through Magic: Visual Awareness as a Source of Visual Metacognition Bias." *Consciousness and Cognition* 65 (2018): 152–68.

Ortega Ortega, Jairo Hernán. "Entrevista con el 'Monstruo.'" *Revista Nova et Vetera* 2, no. 12 (2006). https://www.urosario.edu.co/Revista-Nova-Et-Vetera/Vol-2-Ed-12/Columnistas/Entrevista-con-El-Monstruo%E2%80%9D/.

Ortiz, Darwin. *Strong Magic: Creative Showmanship for the Close-up Magician.* Washington, DC: Richard Kaufman and Alan Greenberg, 1995.

———. *Designing Miracles: Creating the Illusion of Impossibility.* El Dorado Hills, CA: A-1 MagicalMedia, 2006.

Otero-Millán, Jorge, Stephen L. Macknik, Apollo Robbins, Michael McCamy, and Susana Martinez-Conde. "Stronger Misdirection in Curved than in Straight Motion." *Frontiers in Human Neuroscience* 5, no. 133 (2011): 1–4.

Otten, Marte, Yaïr Pinto, Chris L. E. Paffen, Anil K. Seth, and Ryota Kanai. "The Uniformity Illusion: Central Stimuli Can Determine Peripheral Perception." *Psychological Science* 28, no. 1 (January 2017): 56–68.

Parris, Ben A., Gustav Kuhn, Guy A. Mizon, Abdelmalek Benattayallah, and Tim L. Hodgson. "Imaging the Impossible: An fMRI Study of Impossible Causal Relationships in Magic Tricks." *NeuroImage* 45, no. 3 (2009): 1033–39.

Pavese, Cesare. *This Business of Living: Diaries, 1935–1950*. New York: Routledge, 2017.

"Penn and Teller on Broadway." *Talks at Google*, August 3, 2015. https://www.youtube.com/watch?v=5siSa4A9M_Q.

Prevos, Peter. *Perspectives on Magic: Scientific Views on Theatrical Magic*. Kangaroo Flat, Australia: Third Hemisphere Publishing, 2013.

Purves, Dale, and R. Beau Lotto. *Why We See What We Do: An Empirical Theory of Vision*. Sunderland, MA: Sinauer Associates, 2003.

Quian Quiroga, Rodrigo. *Borges and Memory: Encounters with the Human Brain*. Cambridge, MA: MIT Press, 2012.

———. *Qué es la memoria*. Buenos Aires: Paidós, 2015.

———. "Magic and Cognitive Neuroscience." *Current Biology* 26, no. 10 (2016): R390–94.

———. "No Pattern Separation in the Human Hippocampus." *Trends in Cognitive Sciences* 24, no. 12 (2020): 994–1007. https://doi.org/10.1016/j.tics.2020.09.012.

Ramachandran, Vilayanur. *A Brief Tour of Human Consciousness: From Impostor Poodles to Purple Numbers*. New York: Pi Press, 2004.

———. *The Tell-Tale Brain: A Neuroscientist's Quest for What Makes Us Human*. New York: W. W. Norton & Co., 2011.

Randal, Jason. "Misdirection: The Magician's Insurance." *Genii* 40, no. 6 (June 1976): 380–81.

Rayner, Keith, and Monica Castelhano. "Eye Movements." *Scholarpedia* 2, no. 10 (2007): 3649. http://scholarpedia.org/article/Eye_movements.

Raz, Amir, Jay A. Olson, and Gustav Kuhn, eds. "The Psychology of Magic and the Magic of Psychology." *Frontiers in Psychology*, special issue (November 2016). https://www.frontiersin.org/research-topics/2464/the-psychology-of-magic-and-the-magic-of-psychology.

Redelmeier, Donald A., Joel Katz, and Daniel Kahneman. "Memories of Colonoscopy: A Randomized Trial." *Pain* 104, nos. 1–2 (2003): 187–94.

Rensink, Ronald A., and Gustav Kuhn. "A Framework for Using Magic to Study the Mind." *Frontiers in Psychology* 5, no. 1508 (2015): 1–14.

———. "The Possibility of a Science of Magic." *Frontiers in Psychology* 6, no. 1576 (2015): 1–3.

Rensink, Ronald A., J. Kevin O'Regan, and James J. Clark. "To See or Not to See: The Need for Attention to Perceive Changes in Scenes." *Psychological Science* 8, no. 5 (September 1997): 368–73.

Richard. *El mago de los salones o el diablo color de rosa: Nueva colección de juegos de escamoteo, de física y química recreativa, de naipes, magia blanca, etc.*, translated by Juan Traver de Baviñy. Valencia: Librería de Pascual Aguilar, 1875.

Rieiro, Hector, Susana Martinez-Conde, and Stephen L. Macknik. "Perceptual Elements in Penn & Teller's 'Cups and Balls' Magic Trick." *PeerJ* 1 (2013): e19.

Rieznik, Andrés, Lorena Moscovich, Alan Frieiro, Julieta Figini, Rodrigo Catalano, Juan Manuel Garrido, Facundo Álvarez Heduan, Mariano Sigman, and Pablo A. González. "A Massive Experiment on Choice Blindness in Political Decisions: Confidence, Confabulation, and Unconscious Detection of Self-deception." *PLOS ONE* 12, no. 2 (2017): e0171108.

Robert-Houdin, Jean Eugène. *Les secrets de la prestidigitation et de la magie*. Paris: Michel Lévy Frères Editeurs, 1868.

Roediger, Henry L., and Kathleen B. McDermott. "Creating False Memories: Remembering Words Not Presented in Lists." *Journal of Experimental Psychology: Learning, Memory, and Cognition* 21, no. 4 (1995): 803–14.

Roskes, Marieke, Daniel Sligte, Shaul Shalvi, and Carsten K. W. de Dreu. "The Right Side? Under Time Pressure, Approach Motivation Leads to Right-Oriented Bias." *Psychological Science* 22, no. 11 (2011): 1403–7.

Schacter, Daniel L. *The Seven Sins of Memory: How the Mind Forgets and Remembers*. Boston: Houghton Mifflin, 2001.

Schulz, Kathryn. "Fantastic Beasts and How to Rank Them." *The New Yorker*, October 30, 2017. https://www.newyorker.com/magazine/2017/11/06/is-bigfoot-likelier-than-the-loch-ness-monster.

Schweitzer, Richard, and Martin Rolfs. "Intrasaccadic Motion Streaks Jump-Start Gaze Correction." *Science Advances* 7, no. 30 (2021): eabf2218.

Scott, Hannah, Jonathan P. Batten, and Gustav Kuhn. "Why Are You Looking at Me? It's Because I'm Talking, but Mostly Because I'm Staring or Not Doing Much." *Attention, Perception, and Psychophysics* 81, no. 1 (2019): 109–18.

Seneca, Lucius Annaeus. *Seneca's Letters from a Stoic*, translated by Richard Mott Gummere. Mineola, NY: Dover Publications, 2016.

Shalom, Diego E., Maximiliano G. de Sousa Serro, Maximiliano Giaconia, Luis M. Martínez, Andrés Rieznik, and Mariano Sigman. "Choosing in Freedom or Forced to Choose? Introspective Blindness to Psychological Forcing in Stage-Magic." *PLOS ONE* 8, no. 3 (2013): e58254.

Sharot, Tali, Elizabeth A. Martorella, Mauricio R. Delgado, and Elizabeth A. Phelps. "How Personal Experience Modulates the Neural Circuitry of Memories of September 11." *Proceedings of the National Academy of Sciences of the United States of America* 104, no. 1 (2007): 389–94.

Sharpe, Sam H. *Conjurers' Psychological Secrets*. Calgary, Canada: Hades Publications, 1988.

Silvia, Paul, Gil Greengross, Maciej Karwowski, Rebekah Rodriguez, and Sara J. Crasson. "Who Hates Magic? Exploring the Loathing of Legerdemain." *PsyArXiv Preprints*, September 14, 2020. https://psyarxiv.com/mzry6/.

Simons, Daniel J., and Christopher F. Chabris. "Gorillas in Our Midst: Sustained Inattentional Blindness for Dynamic Events." *Perception* 28, no. 9 (1999): 1059–74.

Smith, Tim J. "The Role of Audience Participation and Task Relevance on Change Detection during a Card Trick." *Frontiers in Psychology* 6, no. 13 (February 2015): 1–8.

Smith, Tim J., Peter Lamont, and John M. Henderson. "The Penny Drops: Change Blindness at Fixation." *Perception* 41, no. 4 (2012): 489–92.

———. "Change Blindness in a Dynamic Scene Due to Endogenous Override of Exogenous Attentional Cues." *Perception* 42, no. 8 (2013): 884–86.

Smith, Wally, Frank Dignum, and Liz Sonenberg. "The Construction of Impossibility: A Logic-Based Analysis of Conjuring Tricks." *Frontiers in Psychology* 7, no. 748 (2016): 1–17.

Soon, Chun Siong, Marcel Brass, Hans-Jochen Heinze, and John-Dylan Haynes. "Unconscious Determinants of Free Decisions in the Human Brain." *Nature Neuroscience* 11 (2008): 543–45.

Steinkraus, Warren E. "The Art of Conjuring." *Journal of Aesthetic Education* 13, no. 4 (1979): 17–27.

Stone, Tom. "Vision and Velocity Vectors." *Genii* 74, no. 11 (2011): 40–45.

Strandberg, Thomas, David Sivén, Lars Hall, Petter Johansson, and Philip Pärnamets. "False Beliefs and Confabulation Can Lead to Lasting Changes in Political Attitudes." *Journal of Experimental Psychology: General* 147, no. 9 (2018): 1382–99.

Stromberg, Joseph. "Teller Speaks on the Enduring Appeal of Magic." *Smithsonian Magazine*, February 22, 2012. https://www.smithsonianmag.com/arts-culture/teller-speaks-on-the-enduringappeal-of-magic-97842264/.

Subbotsky, Eugene. "The Belief in Magic in the Age of Science." *SAGE Open* 4, no. 1 (2014): 1–17.

Suchow, Jordan W., and George A. Alvarez. "Motion Silences Awareness of Visual Change." *Current Biology* 21, no. 2 (2011): 140–43.

Tachibana, Ryo, and Jiro Gyoba. "Effects of Different Types of Misdirection on Attention and Detection Performance." *Tohoku Psychologica Folia* 74 (2015): 42–56.

Tachibana, Ryo, and Hideaki Kawabata. "The Effects of Social Misdirection on Magic Tricks: How Deceived and Undeceived Groups Differ." *i-Perception* 5, no. 3 (2014): 143–46.

Tamariz, Juan. "Teoría de la emoción mágica: Apuntes." *Circular de la Escuela Mágica de Madrid* 71, no. 6 (1980): 337–39.

———. "Fundamentos del ilusionismo." In Tamiriz, *Secretos de Magia Potagia*, 33–34. Madrid: Editorial Frakson, 1988.

———. "Sobre la 'pregunta obnubilante.'" *Circular de la Escuela Mágica de Madrid*, 1999.

———. "Por arte de magia: Medio siglo de magia en el mundo." Lecture (audio file). May 26, 2005. Fundación Juan March, Madrid. https://www.march.es/conferencias/anteriores/voz.aspx?p1=2397.

———. *The Five Points in Magic* (1981), corrected edition. Seattle: Hermetic Press, 2007.

———. *The Magic Way: The Theory of False Solutions and the Magic Way* (1988), corrected and augmented edition. Madrid: Frackson, 2011.

———. *The Magic Rainbow*. Seattle: Hermetic Press, 2019.

Tarbell, Harlan. *The Tarbell Course in Magic*, 8 vols. (1927), 15th ed. New York: D. Robbins & Co., 1999.

Thomas, Cyril, and André Didierjean. "Magicians Fix Your Mind: How Unlikely Solutions Block Obvious Ones." *Cognition* 154 (September 2016): 169–73.

———. "The Ball Vanishes in the Air: Can We Blame Representational Momentum?" *Psychonomic Bulletin and Review* 23, no. 6 (2016): 1810–17.

———. "No Need for a Social Cue! A Masked Magician Can Also Trick the Audience in the Vanishing Ball Illusion." *Attention, Perception, and Psychophysics* 78, no. 1 (2016): 21–9.

Thomas, Cyril, André Didierjean, and Serge Nicolas. "Scientific Study of Magic: Binet's Pioneering Approach Based on Observations and Chronophotography." *American Journal of Psychology* 129, no. 3 (Fall 2016): 315–28.

Thomas, Cyril, André Didierjean, François Maquestiaux, and Pascal Gygax. "Does Magic Offer a Cryptozoology Ground for Psychology?" *Review of General Psychology* 19, no. 2 (2015): 117–28.

Tolkien, J.R.R. "On Fairy-Stories." In *The Monsters and the Critics, and Other Essays,* edited by Christopher Tolkien, 109–61. London: George Allen & Unwin, 1983.

Tompkins, Matthew L. "A Science of Magic Bibliography—2021 Update." March 29, 2021. https://www.matt-tompkins.com/blog/2021/3/29/a-science-of-magic-bibliography-2021-update.

Tompkins, Matthew L., Andy T. Woods, and Anne M. Aimola Davies. "The Phantom Vanish Magic Trick: Investigating the Disappearance of a Non-existent Object in a Dynamic Scene." *Frontiers in Psychology* 7, no. 950 (2016): 1–15.

Triplett, Norman. "The Psychology of Conjuring Deceptions." *American Journal of Psychology* 11, no. 4 (1900): 439–510.

Tversky, Amos, and Daniel Kahneman. "Availability: A Heuristic for Judging Frequency and Probability." *Cognitive Psychology* 5, no. 2 (1973): 207–32.

———. "Judgment under Uncertainty: Heuristics and Biases." *Science* 185, no. 4157 (1974): 1124–31.

Van de Cruys, Sander, Johan Wagemans, and Vebjørn Ekroll. "The Put-and-Fetch Ambiguity: How Magicians Exploit the Principle of Exclusive Allocation of Movements to Intentions." *i-Perception* 6, no. 2 (2015): 86–90.

Van Leeuwen, Neil. Book review of *Sleights of Mind: What the Neuroscience of Magic Reveals about Our Brains* by Stephen L. Macknik and Susana Martinez-Conde. *Cognitive Neuropsychiatry* 16, no. 5 (August 2011): 473–78.

Vigen, Tyler. "Spurious Correlations." http://www.tylervigen.com/spurious-correlations (accessed August 12, 2020).

Von Helmholtz, Hermann. "The Facts of Perception" (1878). In *The Selected Writings of Hermann von Helmholtz,* edited by Russell Kahl, 366–408. Middletown, CT: Wesleyan University Press, 1971.

Wagensberg, Jorge. *A más cómo, menos por qué*. Barcelona: Tusquets Editores, 2006.

Wan, Xiaohong, Kang Cheng, and Keiji Tanaka. "Neural Encoding of Opposing Strategy Values in Anterior and Posterior Cingulate Cortex." *Nature Neuroscience* 18 (2015): 752–59.

Ward, Lawrence M. "Attention." *Scholarpedia* 3, no. 10 (2008): 1538. http://www.scholarpedia.org/article/Attention.

Whittlesea, Bruce W. A., and Jason P. Leboe. "The Heuristic Basis of Remembering and Classification: Fluency, Generation, and Resemblance." *Journal of Experimental Psychology: General* 129, no. 1 (2000): 84–106.

Williams, Howard, and Peter W. McOwan. "Magic in the Machine: A Computational Magician's Assistant." *Frontiers in Psychology* 5, no. 1283 (2014): 1–16.

———. "Magic in Pieces: An Analysis of Magic Trick Construction Using Artificial Intelligence as a Design Aid." *Applied Artificial Intelligence* 30, no. 1 (2016): 16–28.

———. "The Magic Words: Using Computers to Uncover Mental Associations for Use in Magic Trick Design." *PLOS ONE* 12, no. 8 (2017): e0181877.

Wilson, Krissy, and Christopher C. French. "Magic and Memory: Using Conjuring to Explore the Effects of Suggestion, Social Influence, and Paranormal Belief on Eyewitness Testimony for an Ostensibly Paranormal Event." *Frontiers in Psychology* 5, no. 1289 (November 2014): 1–9.

Wiseman, Richard, and Emma Greening. "'It's Still Bending': Verbal Suggestion and Alleged Psychokinetic Ability." *British Journal of Psychology* 96 (2005): 115–27.

Wiseman, Richard, Emma Greening, and Matthew Smith. "Belief in the Paranormal and Suggestion in the Seance Room." *British Journal of Psychology* 94 (2003): 285–97.

Wiseman, Richard, and Robert L. Morris. "Recalling Pseudo-Psychic Demonstrations." *British Journal of Psychology* 86, no. 1 (1995): 113–25.

Wiseman, Richard J., and Tamami Nakano. "Blink and You'll Miss It: The Role of Blinking in the Perception of Magic Tricks." *PeerJ* 4 (2016): e1873.

Wonder, Tommy, and Stephen Minch. *The Books of Wonder*, vol. 1. Seattle: Hermetic Press, 1996.

Wright, T. Page, and William W. Larsen. "Mental Effects with Cards—The Beginning of a Series." *Genii* 1, no. 2 (1936): 11–13.

Yao, Richard, Katherine Wood, and Daniel J. Simons. "As if by Magic: An Abrupt Change in Motion Direction Induces Change Blindness." *Psychological Science* 30, no. 3 (March 2019): 436–43.

Yarbus, Alfred L. *Eye Movements and Vision*. New York: Plenum Press, 1967.

Zaghi-Lara, Regina, Miguel Ángel Gea, Jordi Camí, Luis M. Martínez, and Alex Gómez-Marín. "Playing Magic Tricks to Deep Neural Networks Untangles Human Deception." Cornell University, August 20, 2019. https://arxiv.org/abs/1908.07446.

Index

Note: *Italic* page numbers indicate figures.